T0138888

Coded Leadership

Coded Leadership: Developing Scalable Management in an AI-induced Quantum World will assist researchers and industry experts working towards improvising their processes and developing and deploying strategies in an AI-induced world of quantum computing. The book introduces the necessary background to understand the challenges in today's organizational leadership and how artificial intelligence enables learning to be viewed from a probabilistic framework.

Key Features

- Introduction to Quantum Natural Language Processing
- Overview of Leadership and AI
- The Age of Quantum Superiority
- Challenges to Today's Leadership
- AI-induced Strategic Implementation and Organizational Performance

This book serves as a reference for researchers that need to know how AI and quantum can assist in leadership and organizational performance. It book will also be helpful for students that want to learn more about AI and Quantum computing in various business applications.

Coded Leadership

Developing Scalable Management in an AI-induced Quantum World

Edited by
Raul Villamarin Rodriguez
Pinisetti Swami Sairam
Hemachandran K

CRC Press is an imprint of the
Taylor & Francis Group, an **informa** business

First edition published 2023
by CRC Press
6000 Broken Sound Parkway NW, Suite 300, Boca Raton, FL 33487–2742

and by CRC Press
4 Park Square, Milton Park, Abingdon, Oxon, OX14 4RN

CRC Press is an imprint of Taylor & Francis Group, LLC

ISBN: 978-1-032-15554-8 (hbk)
ISBN: 978-1-032-15552-4 (pbk)
ISBN: 978-1-003-24466-0 (ebk)

DOI: 10.1201/9781003244660

Typeset in Minion
by Apex CoVantage, LLC

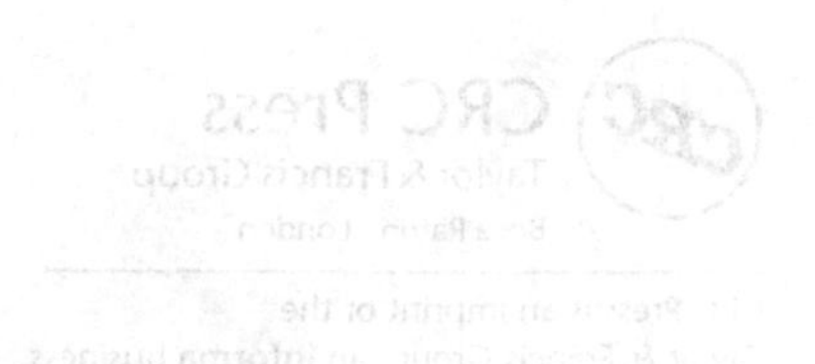

Contents

Preface, vii

Editors, ix

Contributors, xi

CHAPTER 1 ▪ An Introduction to Quantum Natural Language Processing (QNLP) 1

SRINJOY GANGULY, SAI NANDAN MORAPAKULA AND LUIS GERARDO AYALA BERTEL

CHAPTER 2 ▪ When Quantum Meets AI 25

POKALA PRANAY KUMAR AND PARTHAVI SHASTRI

CHAPTER 3 ▪ The Age of Quantum Superiority 35

ASHWIN KUMAAR K, AISHWARYA CHALUVADI AND CYNTHIA JABBOUR SFEIR

CHAPTER 4 ▪ Challenges to Today's Leadership 47

KOLHANDAI YESU, SANJEEV GANGULY AND NARENDRA N. DAS

CHAPTER 5 ▪ A Behavioural Approach to Leadership in Education 61

PREETHA MARY GEORGE, PRINCY SERA RAJAN, BOBBY MAHANTA AND ZOUHOUR EL-ABIAD

CHAPTER 6 ▪ Key Elements That Bind Leadership with AI 71

BHARATH REDDY, PALAK GOEL, VASIREDDY BINDU HASITHA AND ANIL AUDUMBAR PISE

CHAPTER 7 ▪ How Does AI Leadership Affect Strategic Implementation 81

AMOGH S. JAJEE, ANUSHKA JOHARI, DEBDUTTA CHOUDHURY, DAYA SHANKAR, DHEERAJ ANCHURI AND JORGE A. WISE

CHAPTER 8 ▪ The Synergy between AI, Quantum Management, Command and Control 93

AKASH GURRALA, EGUTURI MANJITH KUMAR REDDY AND JUAN R. JARAMILLO

CHAPTER 9 ▪ Quantum Impact on Organizational Performance 103

PREETHAM REDDY GANDAGARI, DEBDUTTA CHOUDHURY, DAYA SHANKAR AND JORGE A. WISE

Index, 119

Preface

ARTIFICIAL INTELLIGENCE IS INFLUENCING THE WAY BUSINESSES ARE performing. Researchers developing AI-based tools and applications draw on a diverse set of disciplines including cognitive science, computer science, statistics, mathematics, philosophy, neuroscience. The need for an AI-induced model in business applications has increased in the post-COVID-19 pandemic. With scalability of business drawing more attention, the criticality in many disciplines is crucial to realize the operational AI capabilities. Creating such systems requires approaches to guide development, deployment and maintenance. A deeper understanding of effective leadership in specific conditions is imperative in the current dynamic environment. Organizations need to develop efforts to find approaches that will enable them to enhance organizational effectiveness.

The chapters in the book aim to provide readers with an AI approach and its impact on the self-development of individuals, technology-enabled leadership and the strategic implementations assisted by artificial intelligence.

We hope our attempt at publishing this book will be beneficial for the student community, industrialists, researchers, their mentors and to all who wish to explore the applications of machine learning. We are greatly thankful to our precious contributors who hail from renowned institutes and industries that have made remarkable contributions by sharing their knowledge for the welfare of the society. We express our sincere, wholehearted thanks to our editorial and production teams for their relentless contribution and for rendering unconditional support to publish this book on time.

Editors

Dr. Raul Villamarin Rodriguez is Vice President at Woxsen University. He holds a PhD in Artificial Intelligence and Robotics Process Automation Applications in Human Resources from San Miguel University, Mexico. Under his leadership, Woxsen University has partnered with 50+ leading universities in 35 countries, across 6 continents. He has been instrumental in bringing international exchange programs and research opportunities with the said global partner universities for students and faculty. His areas of expertise span across the domain of Artificial Intelligence and Quantum Artificial Intelligence, Natural Language Processing, Computer Vision, Robotic Process Automation, Multi-agent Systems, Data Analytics (Big Data), Cybersecurity Management and Knowledge Engineering. Dr. Rodriguez has been part of significant research in his work tenure, especially through RRBM, GRLI and PRME. He has also played vital roles in organizing renowned educational symposiums for institutions like Oxford Brookes University, UK and European Union government bodies in Brussels, Belgium. A young achiever, Dr. Rodriguez was awarded in the Europe India 40 Under 40 Leaders recently, and nominated in Forbes 30 Under 30 Europe 2020. His achievements also highlight him as a registered expert in Artificial Intelligence, Intelligent Systems and Multi-agent Systems at the European Commission. Treading the path of thought leadership in the education domain, he is a member of international bodies such as GRLI Deans and Directors cohort, Harvard Business Review Advisory Council and the Institute for Robotics Process Automation & Artificial Intelligence (IRPA AI). As a staunch knowledge contributor to his field of expertise, he is frequently seen as the keynote speaker and panel moderator at various national and international conferences. He has co-authored two reference books: *New Age Leadership: A Critical Insight* and *Retail Store* and has more than 70 publications to his credit. He is also a weekly contributing writer to various magazines and a notable journal reviewer and associate editor in renowned publications such as IEEE.

Dr. Pinisetti Swami Sairam is Assistant Professor in the School of Business, Woxsen University. Dr. Sairam holds a PhD in Robotics Engineering from UPES, Dehradun; MTech in Robotics Engineering from UPES and BTech in ECE from Vishnu Institute of Technology. He has 8+ years of experience in Robotics, Artificial Intelligence, Embedded Systems and Automation. He has vast experience in implementation, operations and administrative management related to various university accreditation platforms and has authored/co-authored several articles in various SCI/SCOPUS indexed journals and conferences. He started his career as Research Associate with Miranda Automation and Jay Robotix Learning.

Dr. Hemachandran K is Professor of Artificial Intelligence at the School of Business, Woxsen University, Hyderabad, Telangana, India. He has been a passionate teacher for 14 years, with 5 years of research experience. He is a strong educational professional with a flair for science, highly skilled in artificial intelligence and machine learning. After earning a PhD in embedded systems at Dr. M.G.R. Educational and Research Institute, India, he started conducting interdisciplinary research in artificial intelligence. He is open-minded and positive with stupendous peer-reviewed publication records with more than 20 journals and international conference publications. He has served as an effective resource person at various national and international scientific conferences. He has rich research experience in mentoring undergraduate and postgraduate projects. He holds two patents to his credentials. He has life memberships in esteemed professional institutions. His editorial skills have led him to be included as an editorial board member for numerous reputed SCOPUS/SCI journals.

Contributors

Dheeraj Anchuri
School of Business
Woxsen University
Hyderabad, India

Luis Gerardo Ayala Bertel
Faculty of Exact and Natural Science, Mathematics
Cartagena University
Cartagena, Colombia

Aishwarya Chaluvadi
School of Business
Woxsen University
Hyderabad, India

Debdutta Choudhury
School of Business
Woxsen University
Hyderabad, India

Narendra N. Das
Michigan State University
East Lansing, Michigan

Zouhour El-Abiad
Faculty of Economic Sciences and Business
Lebanese University
Beirut, Lebanon

Preetham Reddy Gandagari
School of Business
Woxsen University
Hyderabad, India

Sanjeev Ganguly
Woxsen School of Business
Telangana, India

Srinjoy Ganguly
Fractal Analytics
Gurgaon, India

Preetha Mary George
Educational & Research Institute University
Chennai, India

Palak Goel
School of Technology
Woxsen University
Hyderabad, India

Akash Gurrala
School of Technology
Woxsen University
Hyderabad, India

Vasireddy Bindu Hasitha
School of Technology
Woxsen University
Hyderabad, India

Amogh S. Jajee
School of Business
Woxsen University
Hyderabad, India

Juan R. Jaramillo
Department of Decision Sciences
Adelphi University
Garden City, New York

Anushka Johari
School of Business
Woxsen University
Hyderabad, India

Gabriel Kabanda
California State University
Chico, California

Ashwin Kumaar K
School of Business
Woxsen University
Hyderabad, India

Pokala Pranay Kumar
Woxsen University
Hyderabad, India

Bobby Mahanta
Balurghat College
India

Sai Nandan Morapakula
Department of Electrical & Electronics Engineering
Karunya Institute of Technology & Sciences
Coimbatore, India

Anil Audumbar Pise
School of Computer Science and Applied Mathematics
University of the Witwatersr and
Johannesburg, South Africa

Princy Sera Rajan
Baselios Mathews II College of Engineering
Kerala, India

Bharath Reddy
School of Technology
Woxsen University
Hyderabad, India

Eguturi Manjith Kumar Reddy
School of Technology
Woxsen University
Hyderabad, India

Cynthia Jabbour Sfeir
University of Notre Dame
Lebanon
Zouk Mosbeh, Lebanon

Daya Shankar
School of Business
Woxsen University
Hyderabad, India

Parthavi Shastri
Woxsen University
Hyderabad, India

Jorge A Wise
CETYS University
Baja California, Mexico

Kolhandai Yesu
Woxsen School of Business
Telangana, India

CHAPTER 1

An Introduction to Quantum Natural Language Processing (QNLP)

Srinjoy Ganguly
Sai Nandan Morapakula
Luis Gerardo Ayala Bertel

CONTENTS

1.1	Introduction	2
1.2	Diagrammatic Quantum Theory	2
	1.2.1 Systems and Processes as Wires and Boxes	3
	1.2.2 Process Theory	3
	1.2.3 States, Effects and Numbers	4
	1.2.4 Circuit Diagrams Using Boxes and Wires	4
	1.2.5 Basics of String Diagrams	6
1.3	ZX Calculus	8
1.4	Compositional Grammar	11
1.5	Distributional Words	14
1.6	DisCoCat Algorithm	15
1.7	Converting Sentences into Circuits Using ZX Calculus	16
1.8	Process of Scalable QNLP Using CCG Concepts	20
References		23

DOI: 10.1201/9781003244660-1

1.1 INTRODUCTION

Quantum Natural Language Processing (QNLP) is an emerging field where we hope to achieve significant advantages by using a quantum computer to process languages. This field is in a very nascent stage but still significant progress has been achieved to prove that QNLP concepts do work in practice when they are executed on real quantum hardware. The most beautiful aspect of QNLP is that it connects abstract mathematical concepts such as category theory (especially monoidal category) to real-world practical implementation. With the advent of large amounts of data, classical computers are gradually losing their processing power and as a consequence it takes a long time to process large chunks of data. Quantum computers bring a promise to handle large chunks of data such as text in an efficient manner.

This chapter introduces the concepts underlying QNLP and we tend to keep things very diagrammatic. The reason behind this is because QNLP itself is completely pictorial and the pictures have their own mathematical identity which we are not going to cover here but is a worthy point to note. Another aspect of QNLP is that it connects two different fields—categorical quantum mechanics and linguistics together to prove that QNLP is actually quantum-native, which means NLP should be performed on quantum computers not on classical.

This chapter starts with an introduction to diagrammatic quantum theory and then moves on to explain the ZX calculus, which is a graphical language used to make quantum circuits. Then we explain the DisCoCat (Distributional Compositional Categorical) algorithm which is at the heart of the QNLP algorithm and we give a short introduction to converting diagrams into quantum circuits. Finally, we touch briefly on CCG (Combinatory Categorial Grammar) which paves a way for scalable QNLP in the near future.

1.2 DIAGRAMMATIC QUANTUM THEORY

The origin of diagrammatic quantum theory goes back to the early 2000s. The very first paper on diagrammatic reasoning was written by Bob Coecke in 2003 and then the field of categorical quantum mechanics came in 2005. The concepts explained in this section are based on Coecke et al. [1] and are just the diagrammatic representation of some category theory concepts, which can be used for quantum theory. The most elegant aspect of this diagrammatic nature is that you do not need to dive deep into abstract mathematical definitions of category theory.

1.2.1 Systems and Processes as Wires and Boxes

We can say that objects are the systems and morphisms are the processes. To denote these systems and morphisms diagrammatically, we use boxes and labelled wires. Therefore, labelled wires are known as the systems and the boxes are known as the processes. Figure 1.1 is the representation of systems and processes.

The bottom wires are the inputs to a box (process) and the top wires are the outputs of a process. Those labelled wires are the system types and as can be seen clearly, the types A & D of the right processes match that of left process output and input, respectively.

From Figure 1.2 we can come to the conclusion that systems or processes can only be combined if their types match (closely observe the last diagram).

1.2.2 Process Theory

For the proper interpretation of these boxes and wires, it is important to understand process theory because there are various kinds of processes such as biological, chemical, etc.

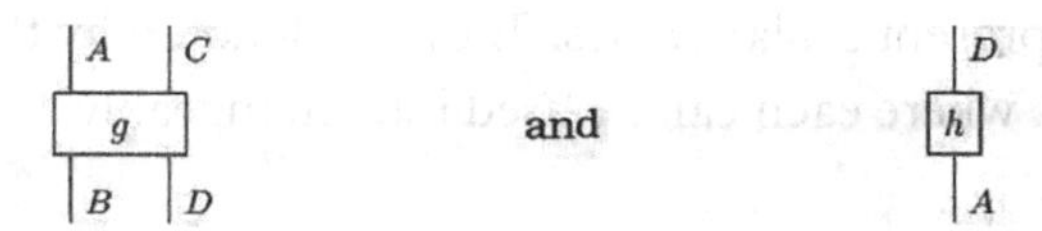

FIGURE 1.1 Representation of systems and processes.

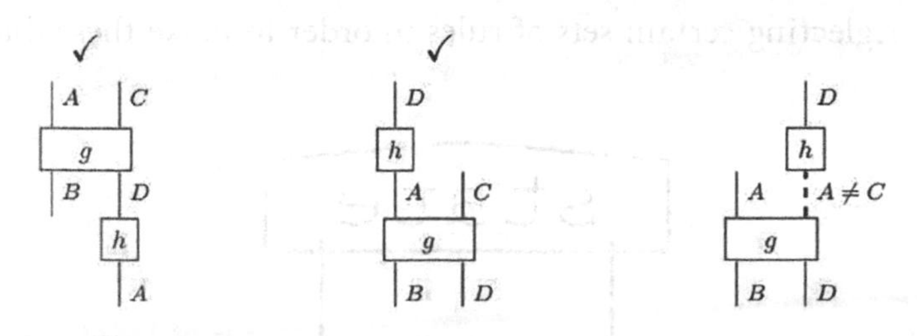

FIGURE 1.2 Combination of systems or processes.

A process theory consists of:

(i) A collection T of system types represented by wires,

(ii) A collection P of processes represented by boxes,

(iii) A means of 'wiring processes together'.

1.2.3 States, Effects and Numbers

These are the special kind of processes in quantum theory.

1) *States*: The processes which do not possess any inputs are called states. States are also called 'Dirac Ket' which is represented by |state >. Figure 1.3 is a diagrammatic representation of state.

   States are used to represent words for QNLP tasks.

2) *Effects*: The processes which do not possess any outputs to them and which are considered dual of states are called effects. Effects are also called 'Dirac Bra' which is denoted by effect|. Figure 1.4 is the diagrammatic representation of effect.

   Effects are used as quantum processes for QNLP tasks.

3) *Numbers*: The processes without any inputs or outputs. They are used to represent scalar values. They are denoted by the diagrams in Figure 1.5, where each can be used interchangeably.

1.2.4 Circuit Diagrams Using Boxes and Wires

As discussed above, the boxes and the wires can be combined together without neglecting certain sets of rules in order to make these diagrams

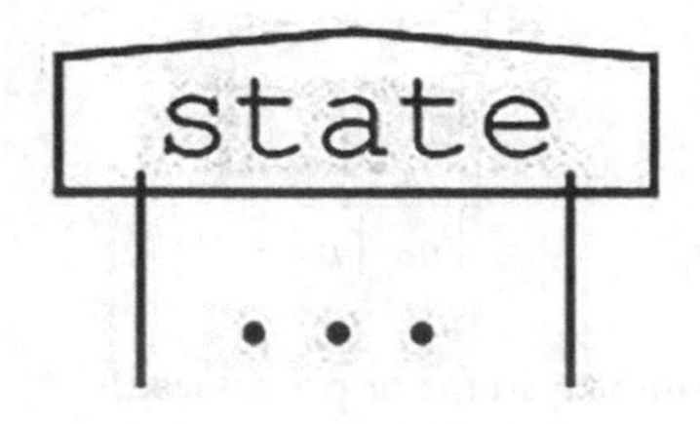

FIGURE 1.3 Diagrammatic representation of state.

FIGURE 1.4 Diagrammatic representation of effect.

FIGURE 1.5 Diagrammatic representation of numbers.

convert into a circuit diagram. These three rules are based on composition and directed cycles:

(i) Parallel composition

(ii) Sequential composition

(iii) No directed cycles

1) *Parallel Composition*: It is denoted by '⊗' symbol and consists of placing the pair of diagrams side by side which means that the processes happen independently of each other. This is shown in Figure 1.6.

2) *Sequential Composition*: It is denoted by 'o' and consists of connecting the outputs of the first diagram with the inputs of the second diagram. This is shown in Figure 1.7.

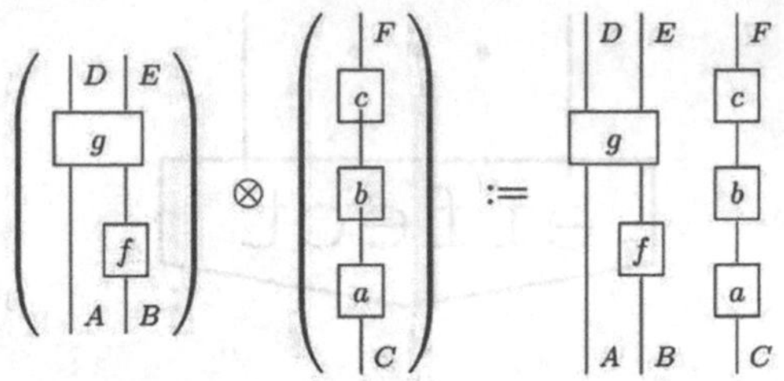

FIGURE 1.6 Diagrammatic representation of parallel composition.

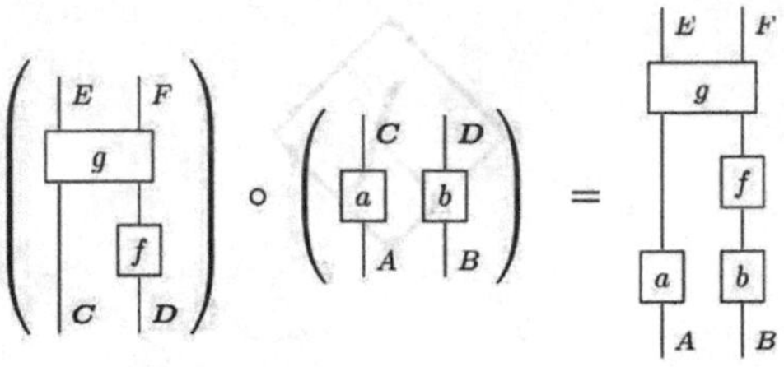

FIGURE 1.7 Diagrammatic representation of sequential composition.

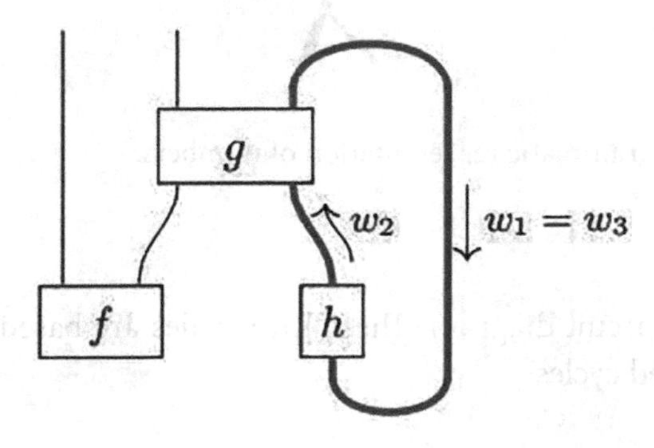

FIGURE 1.8 Diagrammatic representation of no directed cycles.

3) *No Directed Cycles*: No directed cycles should be present in the diagram. The diagram in Figure 1.8 shows a directed cycle and these kinds of cycles are not allowed in a circuit diagram.

1.2.5 Basics of String Diagrams

In the previous sections, diagrammatic calculus was introduced. With that covered, it is now helpful to understand a few more diagrams which bring the most important quantum phenomena in a pictorial fashion, namely—entanglement (a phenomena in which quantum states cannot be described independently). As soon as we get familiar with string diagrams basics, it becomes easy to comprehend the QNLP diagrams for the DisCoCat model.

So what are string diagrams? The answer to this question can be given in two ways.

1. Diagrams which consist of boxes, wires and additionally inputs being connected to inputs and outputs being connected to outputs forming cycles in the diagram which is shown in Figure 1.9.

   In the previous section we said that these cycles were not allowed in the circuit diagrams. Allowing them now gives the diagrams better functionality to define more complex operations.

2. Circuit diagrams which contain special states and special effects, namely the cup state and cap effect. This is shown in Figure 1.10.

   It is important to note that for each cup and cap, the diagrams in Figure 1.11 hold true.

   Also from the diagram in Figure 1.11, the diagram in Figure 1.12 makes sense.

It can be carefully observed that the crossings present in the wires are yanked to get the caps and cups. It is because of this, the diagram is also known as a yanking equation.

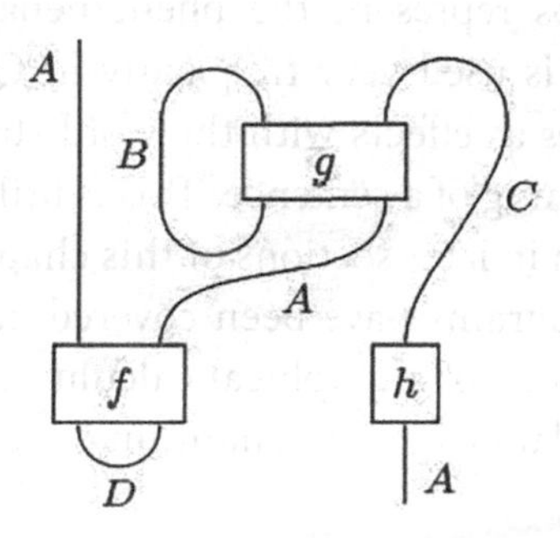

FIGURE 1.9 Diagrammatic representation of cycles.

FIGURE 1.10 Diagrammatic representation of cup state and cap effect.

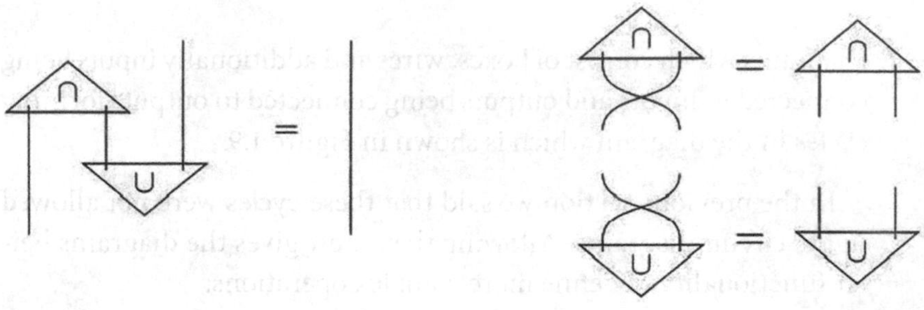

FIGURE 1.11 Rules of cups and caps.

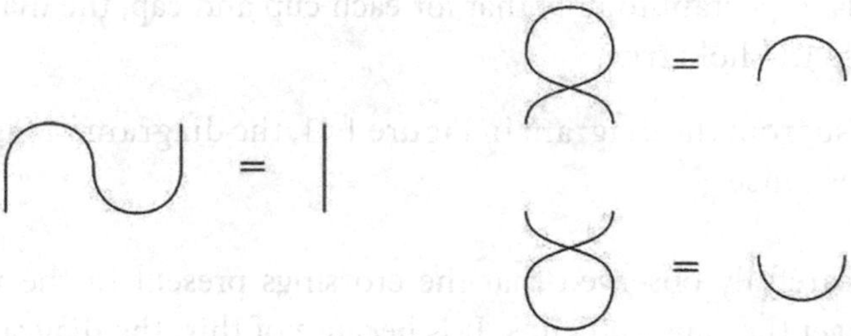

FIGURE 1.12 Another way of representing Figure 1.11.

These cups and caps represent the phenomena of entanglement or entangling effect. This is used very frequently in QNLP models to connect quantum processes as effects with the word states to compose them together to form a meaning of a sentence. This is utilized by the DisCoCat algorithm which is seen in later sections of this chapter.

Now since string diagrams have been covered, it is also important to learn some basic notations of a graphical calculus called the ZX calculus which helps to convert language into quantum circuits.

1.3 ZX CALCULUS

This section is based on the work from Yeung and Kartsaklis [2]. What is ZX calculus? If you are thinking about the same question, don't worry!

We've got you covered. ZX calculus is a graphical language which is used to represent quantum circuit diagrams as liner maps between qubits. In this section, we try to use good intuition on ZX calculus.

The ZX diagrams consist of diagrammatic rewrite rules which assist in the reasoning of linear maps between qubits and are used to convert string diagrams into quantum circuit diagrams.

ZX diagrams are generated by some basic generators which are called spiders. There are two types of spiders: 'White Dot' and 'Grey Dot'. Inputs are the lines on the left and outputs are on the right.

The 'white dot' is known as the Z spider. It can be seen from Figure 1.13 that since it looks like spiders, you know from where they got that name.

The Z spiders denote the Z basis or the computational basis |0> and |1>. It is worth noting that due to presence of α inside the spider, it becomes a 'decorated' or 'phase' spider. This means α is the phase and therefore a decorated Z spider is the same as the RZ gate for 1 input and 1 output as seen in Figure 1.14.

And we know that the RZ gate does custom rotations around the Z axis of the Bloch sphere taking α as the angle of rotation around the Bloch sphere.

The 'grey dot' is known as the X spider. The X spiders denote the X basis or the diagonal basis states |+> and |->. The diagram in Figure 1.15 represents a phased X spider.

The X spider with [1 input and 1 output] corresponds to the RX gate as shown in Figure 1.16.

$$:= \quad |0\cdots0\rangle\langle0\cdots0| + e^{i\alpha}\,|1\cdots1\rangle\langle1\cdots1|\,,$$

FIGURE 1.13 Diagrammatic representation of Z spider.

$$= \quad |0\rangle\langle0| + e^{i\alpha}\,|1\rangle\langle1| \quad = \quad \begin{pmatrix}1 & 0\\ 0 & 0\end{pmatrix} + \begin{pmatrix}0 & 0\\ 0 & e^{i\alpha}\end{pmatrix} \quad = \quad \begin{pmatrix}1 & 0\\ 0 & e^{i\alpha}\end{pmatrix}$$

FIGURE 1.14 Representation of Z spider in matrix form.

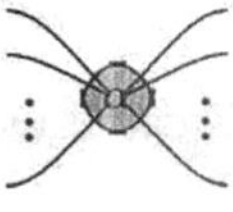

$$:= \quad |+\cdots+\rangle\langle+\cdots+| + e^{i\alpha}\,|-\cdots-\rangle\langle-\cdots-|\,.$$

FIGURE 1.15 Diagrammatic representation of X spider.

$$= |+\rangle\langle+| + e^{i\alpha}|-\rangle\langle-| = \frac{1}{2}\begin{pmatrix}1 & 1\\ 1 & 1\end{pmatrix} + \frac{1}{2}e^{i\alpha}\begin{pmatrix}1 & -1\\ -1 & 1\end{pmatrix} = \frac{1}{2}\begin{pmatrix}1+e^{i\alpha} & 1-e^{i\alpha}\\ 1-e^{i\alpha} & 1+e^{i\alpha}\end{pmatrix}$$

FIGURE 1.16 Representation of X spiders in matrix form.

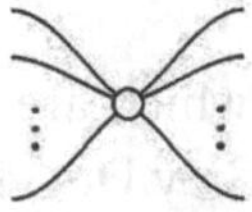

$$= |0\cdots0\rangle\langle0\cdots0| + |1\cdots1\rangle\langle1\cdots1|$$

FIGURE 1.17 Diagrammatic representation of Z spiders when $\alpha = \pi$.

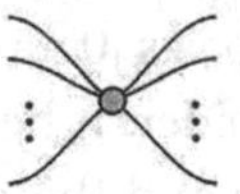

$$= |+\cdots+\rangle\langle+\cdots+| + |-\cdots-\rangle\langle-\cdots-|$$

FIGURE 1.18 Diagrammatic representation of X spiders when $\alpha = \pi$.

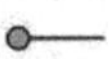

$$= |+\rangle + |-\rangle = \sqrt{2}\,|0\rangle \qquad = |0\rangle + |1\rangle = \sqrt{2}\,|+\rangle$$
$$= |+\rangle - |-\rangle = \sqrt{2}\,|1\rangle \qquad = |0\rangle - |1\rangle = \sqrt{2}\,|-\rangle$$ 

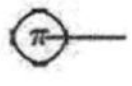

FIGURE 1.19 Diagrammatic representation of Z and X gates in ZX calculus.

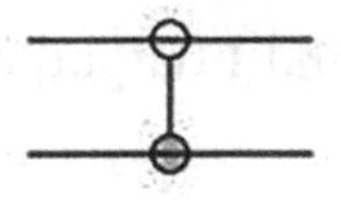

FIGURE 1.20 Diagrammatic representation of CNOT gate in ZX calculus.

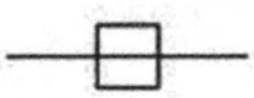

$$= e^{-i\frac{\pi}{4}}$$

FIGURE 1.21 Diagrammatic representation of Hadamard gate in ZX calculus.

Let us assume that $\alpha = \pi$ for above given examples in Figures 1.17 and 1.18, and see how Z spiders and X spiders look (Figure 1.19). Here we go,

- Z spiders
- X spiders

I have a question here. What do quantum gates look like in ZX calculus? Interesting? Let us have a look at them.

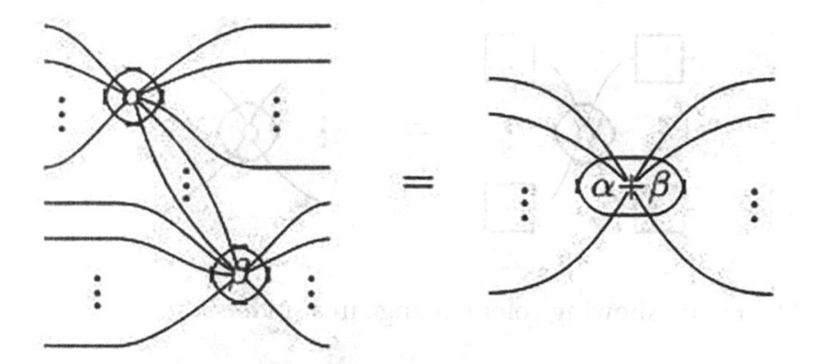

FIGURE 1.22 Representation of a two spider fuse.

Pauli Z and X basis states can be denoted by the following spiders.

- CNOT gate in ZX calculus (Figure 1.20).
- Hadamard gate can be constructed as seen in Figure 1.21.

Until now, we have covered ZX diagrams; now let us see some of the rewrite rules in ZX calculus.

1. Spider Fusion

   Adjacent spiders of the same colour fuse and their phases add which can be seen in Figure 1.22.

2. Colour Change

   The two spiders are related to each other by Hadamard gates. This can also be seen as a rule for commuting a Hadamard gate through a spider. The colour change can be seen in Figure 1.23.

With the help of the basic rewrite rules, there are certain circuit identities for ZX calculus which can be used for analysing quantum circuits and converting string diagrams into quantum circuits. Some of them are given in Figures 1.24 and 1.25.

With the ZX calculus covered, we are now ready to dive into the main component of the QNLP concept called the DisCoCat algorithm.

1.4 COMPOSITIONAL GRAMMAR

Now we are gradually getting into the core concepts of QNLP. This section and the next two sections after this are based on Coecke and Kissinger [3].

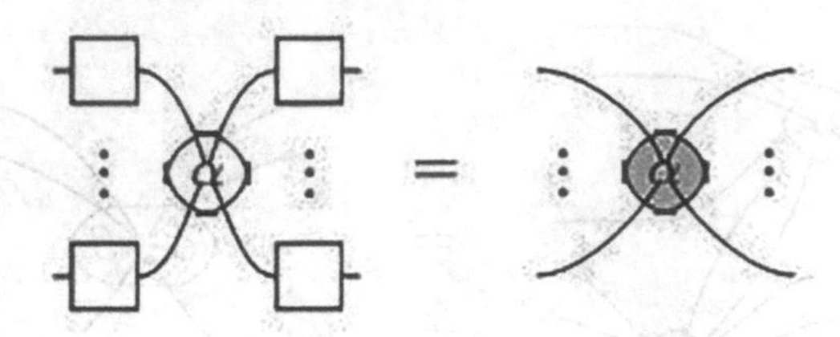

FIGURE 1.23 Figure showing colour change in a spider.

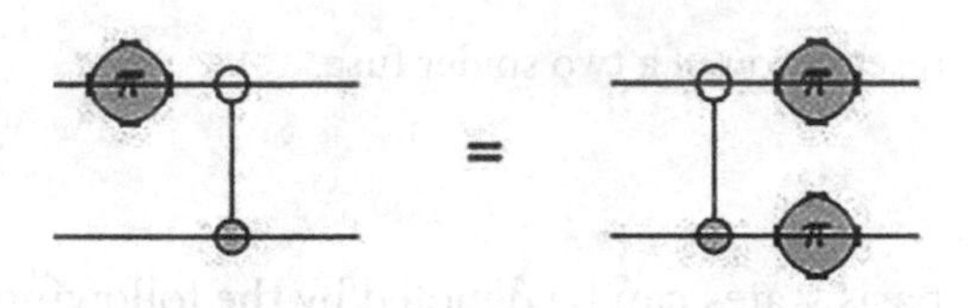

FIGURE 1.24 Circuit identity.

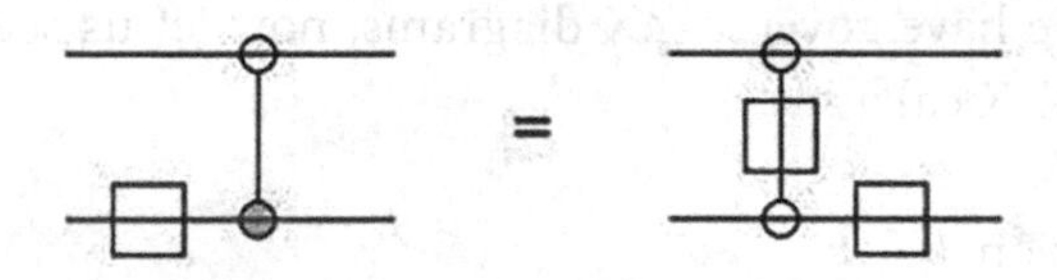

FIGURE 1.25 Circuit identity.

Before getting started with the compositionality of grammar, let us get ourselves familiar with pregroups.

PREGROUP: A Pregroup (P, $\leq$, $\cdot$, 1, $(-)^l$, $(-)^r$) is a partially ordered monoid in which each element p $\in$ P has a left adjoint p^l and a right adjoint p^r, i.e. the following hold: $p^l \cdot p \leq 1 \leq p \cdot p^l$ and $p \cdot p^r \leq 1 \leq p^r \cdot p$.

Some of the important properties of pregroups are

- Adjoints are unique
- Adjoints are order reversing: $p \leq q \Rightarrow q^r \leq p^r$ and $q^l \leq p^l$
- The unit is a self-adjoint: $1^l = 1 = 1^r$
- Multiplication is a self-adjoint: $(p \cdot q)^r = q^r \cdot p^r$
- Opposite adjoints annihilate each other: $(p^l)^r = p$
- Same adjoints iterate

Even if you don't understand what a pregroup actually is, take it easy (if you are not interested) but try to observe the properties well enough. The pregroup properties will take you through.

To understand a pregroup applied to a sentence, some notations regarding grammar are necessary. The noun is denoted by *n* and a sentence is denoted by *s*. An example is shown in Figure 1.26 to make things clear.

We can see from Figure 1.26 that Virat and Burger are nouns and the word eats is a transitive verb. If the juxtaposition of these two types of words in a sentence reduces to the basic type *s* the sentence is said to be a grammatically well-typed sentence. This means that sentence which is said to be grammatically well-typed should follow and satisfy the equation in Figure 1.27.

The diagrammatic representation of the above cancellation is shown in Figure 1.28 where the cups help provide structure to a sentence i.e. it symbolizes the grammar compositionality, where it composes each of the words in the sentence.

Therefore, combining the above diagram reductions and representing each of the words as states, the diagram in Figure 1.29 makes sense.

Virat	**eats**	**Burger**
n	$(n^r s n^l)$	n

FIGURE 1.26 Figure showing some notations regarding grammar.

$$n(n^r s n^l)n \rightarrow 1 s n^l n \rightarrow 1 s 1 \rightarrow s$$

FIGURE 1.27 Grammatically well-typed sentence.

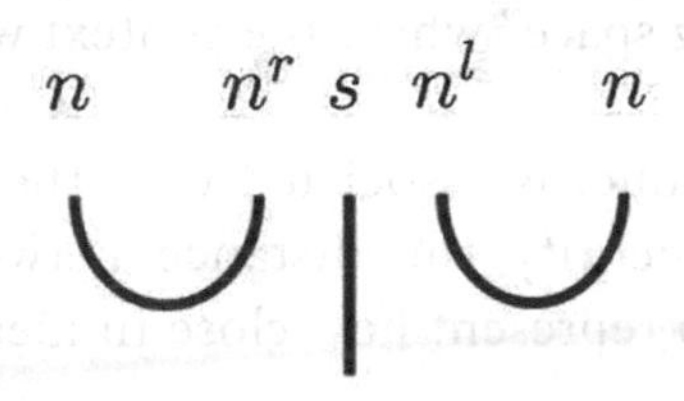

FIGURE 1.28 Diagrammatic representation of Figure 1. 27.

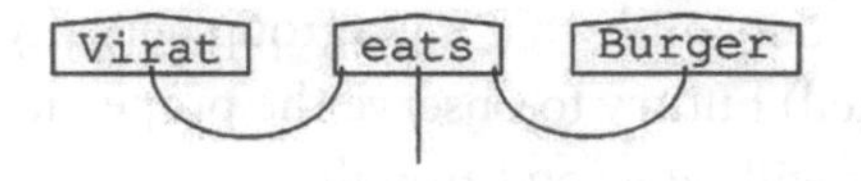

FIGURE 1.29 Diagrammatic representation of a sentence.

From the above examples it is clearly seen that pregroup grammar is a compositional grammar model which can give meaning to large syntactic units present in the sentence. To put it even simpler, the composition formed by the cups and wires is the grammatical structure.

Since pregroup grammars have a natural tensor structure, they are therefore said to be quantum native. Because of this quantum native nature that it becomes evident that Natural Language Processing (NLP) is supposed to give better advantages and results if run on a quantum computer and this is known as Quantum Natural Language Processing (QNLP).

1.5 DISTRIBUTIONAL WORDS

The compositional grammar model—pregroup grammar, which is explained in the previous section, is incomplete unless we have a model to embed the meaning of various words present in a sentence. In this section let us see how it can be done.

Distributional word representation depends on the vector space models for meaning. It consists of finding the meaning of a word according to the context in which it appears. For example, if we have the words vegetables and fruits then they can appear as subject for the context words cook, eat and healthy. Consider a corpus where these words appear five times, seven times and nine times respectively. Then the vector representation for the word shirt becomes [5, 7, 9]. This means that the meaning of the words can be easily depicted in a high-dimensional "meaning space" where the context words act as orthogonal basis vectors.

Since the inner product is associated with the vector space model, it becomes easy to identify the distance between different words which in turn helps to represent how close in meaning is one word to another.

If there are three words such as Mike, red and eats then they can be represented using the diagrams in Figures 1.30 and 1.31 which depict the states in the same vector space.

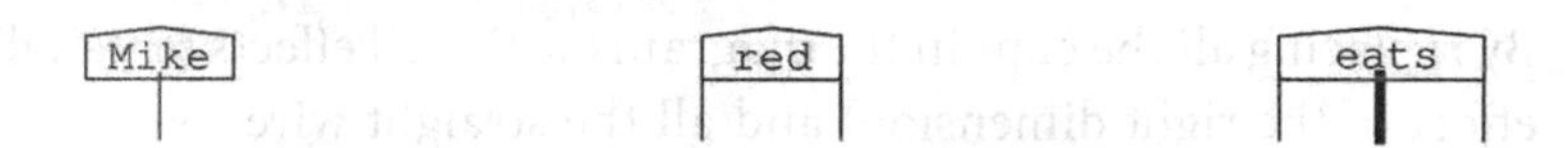

FIGURE 1.30 Diagram showing dimensional spaces.

The black bold line in the middle for the word 'eats' denotes that it lives a very high-dimensional space and needs to be reduced in order to carry out computations on today's hardware.

Now since you have seen compositional nature of grammar and distributional nature of words, they can be now combined to form the DisCoCat algorithm which is explained in the next section.

1.6 DisCoCat ALGORITHM

In the previous sections we covered the compositional nature of grammar and distributional nature of words. In this section, we combine those previous concepts together to introduce the algorithm called DisCoCat (Distributional Compositional Categorical) which is the core component of QNLP and it is a compositional model which is able to provide meaning of a whole (sentence) by combining (entangling) all the parts (words). This is also called the bottom-up approach.

Here are the steps for the DisCoCat algorithm:

1. Assume that there are words wl · . . . · wN, which grammatically make sense, the meanings of the parts are given, and the spaces in which these meanings of parts live with respect to the grammatical type. For example Figure 1.31

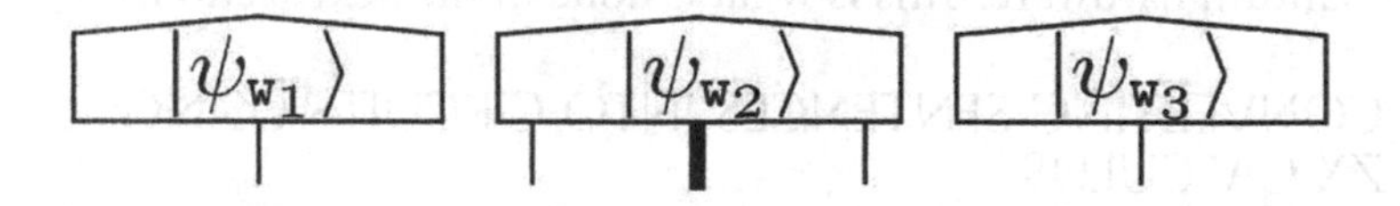

FIGURE 1.31 Diagrammatic representation of words.

2. The wire diagram representing grammar (entangling effect) is established, for example Figure 1.32

FIGURE 1.32 Diagram representing grammar.

3. By replacing all the cups in the diagrams with Bell effects (entangling effect of the right dimension) and all the straight wires with identities, a quantum process is produced as seen in Figure 1.33.

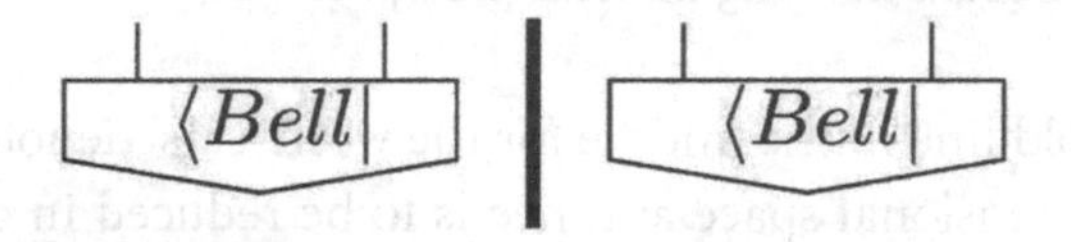

FIGURE 1.33 Diagrammatic representation of a quantum process.

4. We apply this quantum process to the meanings of the parts as shown in Figure 1.34.

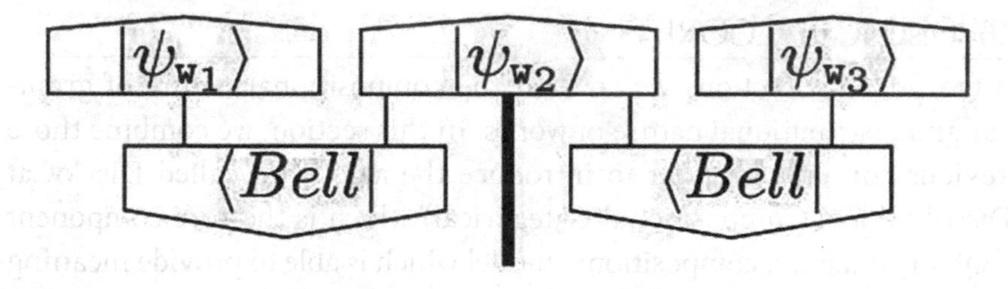

FIGURE 1.34 Quantum process applied to meanings of parts.

The result obtained is the meaning of the whole sentence. Here we have only considered a simple case of subject-verb-object, but, string diagrams of more complex sentences can be formed by using the rules given by Coecke [5].

The resulting string diagram which we obtained from the DisCoCat algorithm cannot be run directly on a quantum computer. A quantum circuit of the string diagram will be required in order to run this sentence on real quantum hardware. This is what is done in the next section.

1.7 CONVERTING SENTENCES INTO CIRCUITS USING ZX CALCULUS

The ZX diagrams covered in this section are based on Coecke and Kissinger [3] which provides more references and details on the process of converting string diagrams to quantum circuits.

Let us tackle one of the most important and interesting question in this section: 'How are we going to tell or convert the language we understand into the language of quantum computers?' We have generated the string diagrams using the DisCoCat algorithm in the above section, but the language we speak and understand is not the same language quantum computers speak (*SCIFI Movie Feels*) and understand, right?

In this section using ZX calculus, let us see how to convert the string diagrams into quantum circuits (the only possible way to talk to quantum computers) by which quantum computers can learn the grammar and meaning of the sentences.

The major constraint here is with the dimensions of meaning spaces for NLP which can reach up to very large numbers (10^{24}) which are not feasible to implement in any hardware device we currently have. But don't lose hope yet! We have our backs covered by the internal wiring system, which can reduce the dimensions to around (10^{6}) and which are very feasible to work with. The internal wiring system is provided as Figure 1.35.

The dimensionality reduction of the word 'eats' is shown diagrammatically in Figure 1.36. We can see from the Figure 1.36 that the bold line (high dimension) is removed and only two lines (low dimension) remain.

Figure 1.37 tells us that the wiring system allows the sliding of spiders towards the nouns Virat and Burger.

Based upon the ZX calculus rewrite rules, the previous diagram in Figure 1.37 can be applied with the following technique of spiders shown in figure 1.38.

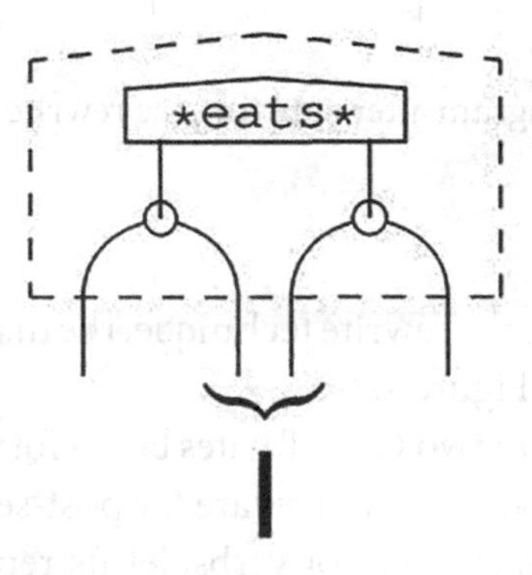

FIGURE 1.35 Figure showing internal wiring of a system.

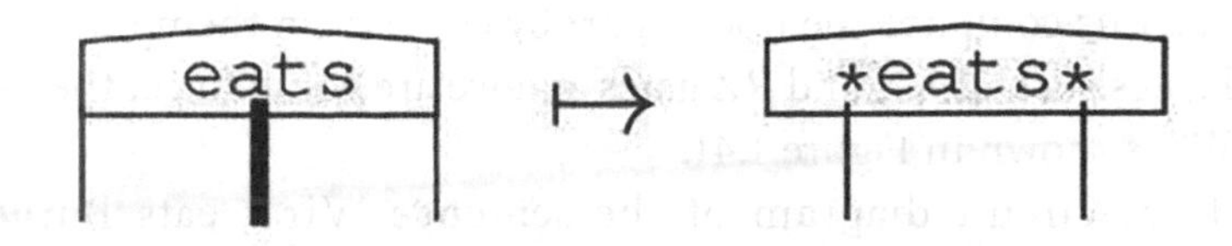

FIGURE 1.36 Dimensionality reduction of the word eats.

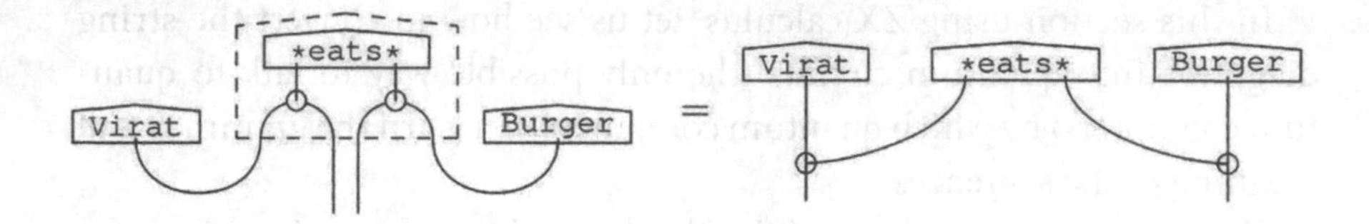

FIGURE 1.37 Figure showing the sliding of spiders.

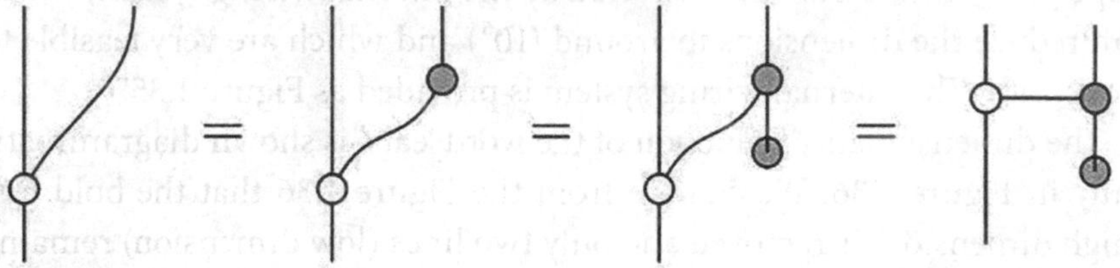

FIGURE 1.38 Representation of Figure 1.37 using ZX calculus rewrite rules.

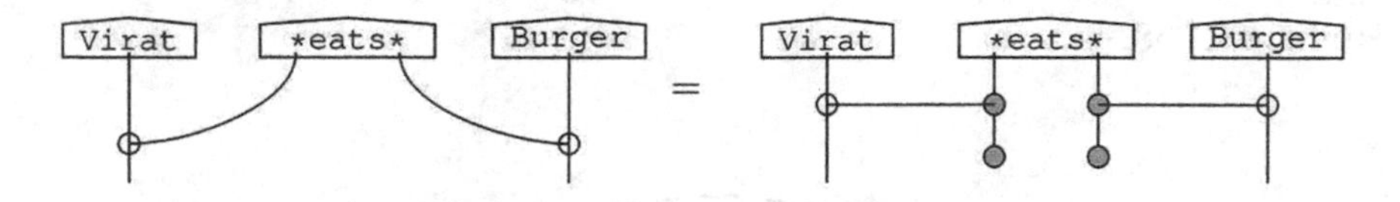

FIGURE 1.39 Output diagram after applying the rewrite rules.

By carefully applying the rewrite technique, the diagram that is obtained is given below given in Figure 1.39.

It is seen that there are two CNOT gates being formed when the spiders are pulled out and the last two spiders are for post-selected measurement.

Coming to the representation of verbs, let us represent the verbs by a unitary gate U so that we can easily represent a large space consisting different verbs each with different values of α, β and ϒ (Figure 1.40).

Before having a look at the complete quantum circuit of a particular sentence, let us complete the nouns part by converting them into parameterized gates such as RX and RZ gates and we are going to get the following circuit as shown in Figure 1.41.

The final circuit diagram of the sentence 'Virat eats Burger' is shown in Figure 1.42, which clearly marks the subject-verb-object in the circuit.

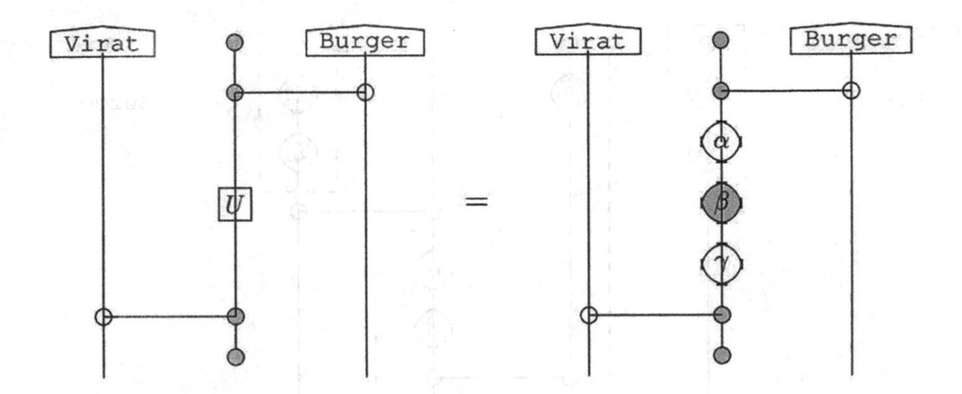

FIGURE 1.40 Representation of verbs.

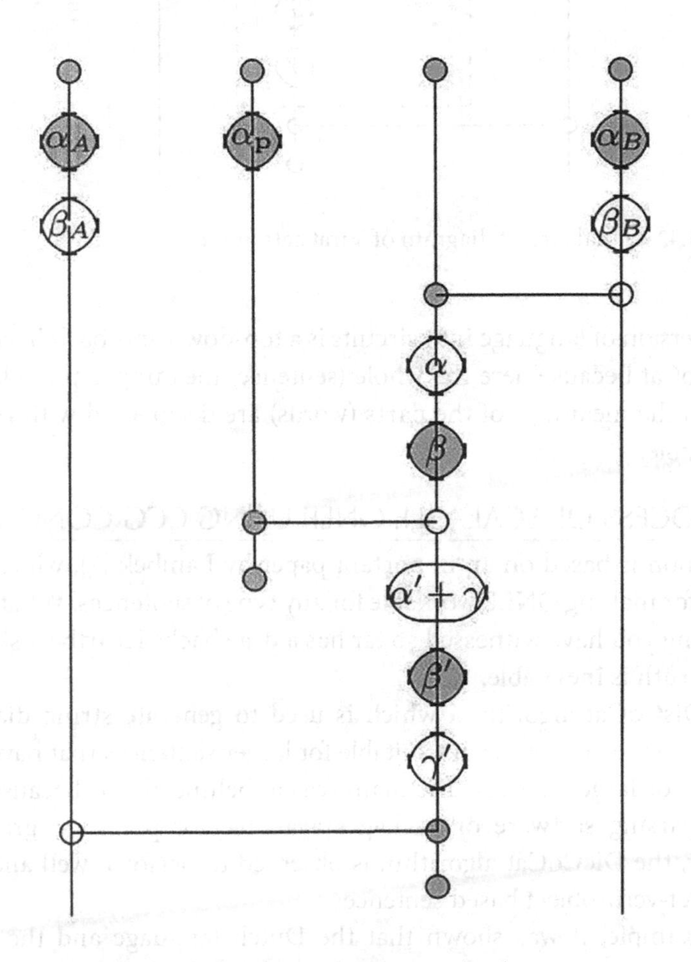

FIGURE 1.41 Conversion of nouns into parameterized gates.

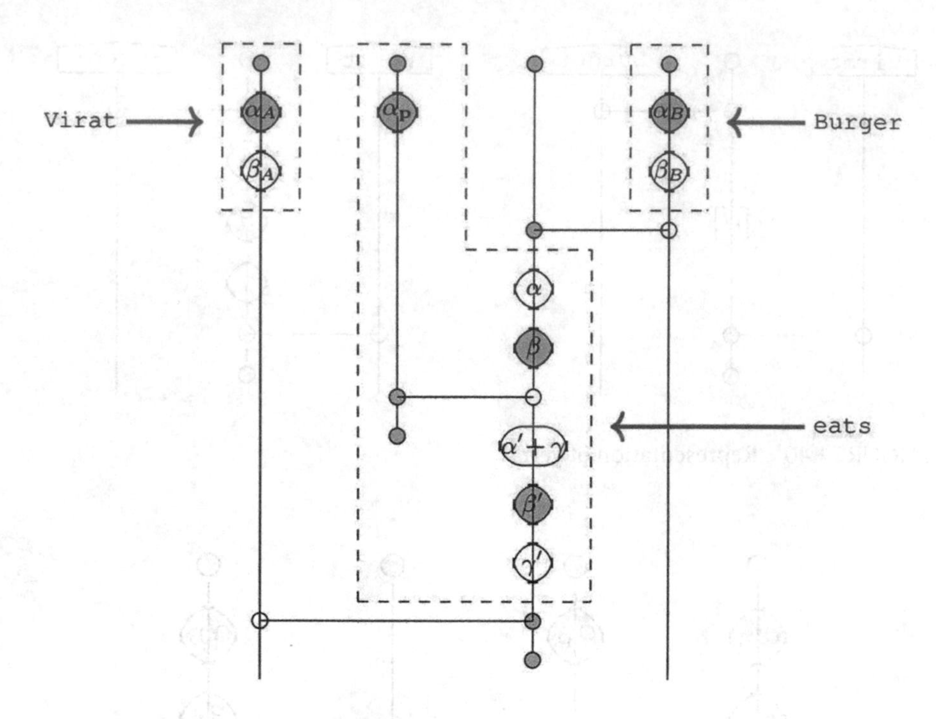

FIGURE 1.42 Final circuit diagram of Virat eats burger.

Conversion of language into circuits is a top-down approach in contrast to DisCoCat because here the whole (sentence) meaning is provided first and then the meanings of the parts (words) are deciphered with the help of the whole.

1.8 PROCESS OF SCALABLE QNLP USING CCG CONCEPTS

This section is based on an important paper by Lambek [7], which paves the way for making QNLP workable for any type of sentences. What if I say everything you have witnessed so far has a drawback? Heartbeat skipped! But the truth is inevitable.

The DisCoCat algorithm, which is used to generate string diagrams that we have seen above is not suitable for larger sentences that have more grammar or large datasets. The main reason behind this is because there are no parsing software or packages available for pregroup grammar. However, the DicCoCat algorithm is observed to perform well and good for subject-verb-object based sentences.

For example, it was shown that the Dutch language and the Swiss-German languages share cross-serial dependencies.

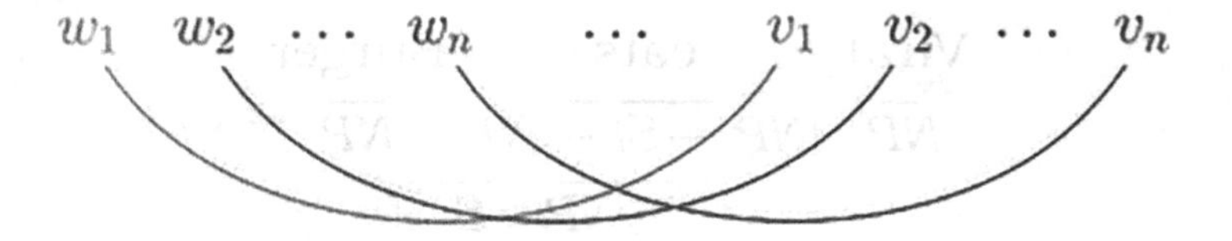

FIGURE 1.43 Figure showing the cross-serial dependencies of Swiss-German languages.

The image in Figure 1.43 from Wikipedia shows the example of cross-serial dependencies which the words w's and v's overlap with each other and are therefore not context-free. This aspect of cross-serial dependencies is not covered by the DisCoCat algorithm.

The Categorial Grammar first introduced where certain syntactic constituent functions are applied to the lower order arguments, e.g. an intransitive verb will have a type S\NP. The S\NP means that a Noun Phrase (NP) needs to be on the left in order to get a Sentence (S). The direction of the slash determines the position of the argument. In general, for type X\Y, it takes type Y to be on the right to return type X and similarly for X/Y expects to return type X when type Y is on the left side (Figure 1.44).

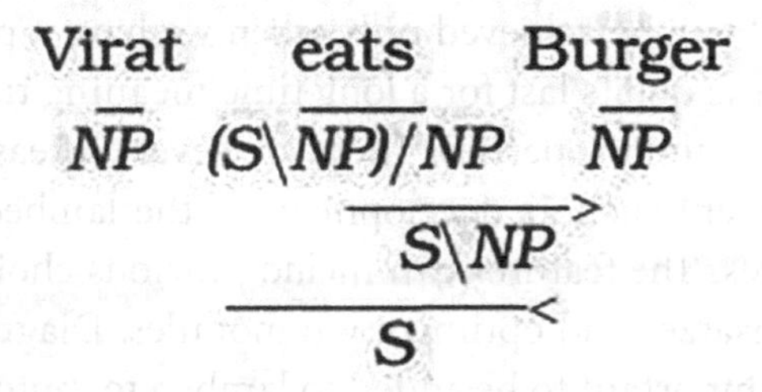

FIGURE 1.44 Representation of categorical grammar in a sentence.

In the above sentence, the transitive verb 'eats' has the type (S\NP)/NP which makes sense because, the sentence can be returned only when there are two noun phrases on the left and right side, which is true i.e. Virat and Burger are those two noun phrases.

Replacing X/Y with X ← Y and X\Y by Y → X the notation in Figure 1.45 is achieved.

The notation displayed above Figure 1.45 is the basic categorial grammar formalism and when this formalism is added with a few more rules such as type-raising and composition then this categorial grammar becomes a

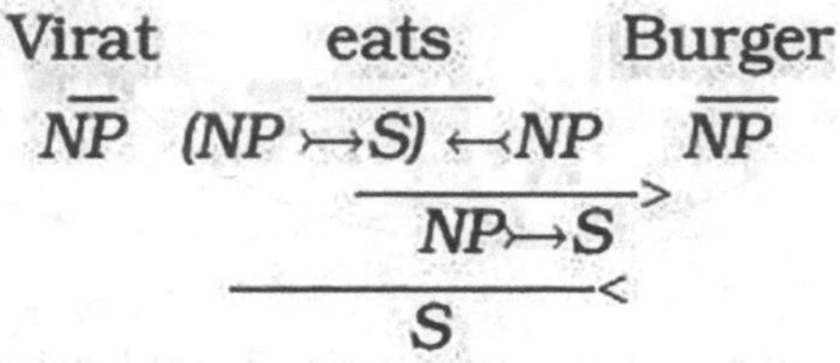

Virat eats Burger

$\overline{NP}$ $\overline{(NP \rightarrowtail S) \leftarrowtail NP}$ $\overline{NP}$

$\underline{\qquad\qquad}>$

$NP \rightarrowtail S$

$\underline{\qquad\qquad}<$

S

FIGURE 1.45 Figure showing the notation after replacing X/Y with X ← Y and X\Y by Y → X.

combinatory categorial grammar (CCG) which provides greater flexibility and power to combine different syntactic constituents of a sentence in any order.

The recently released QNLP toolkit by Cambridge Quantum called lambeq [7] provides a CCG parser, which creates the CCG derivation of the sentence provided and then converts it into a string diagram. After some rewriting of the string diagrams, they are converted into a quantum circuit. Because of this CCG parser and other interesting features, lambeq provides a way towards scalable QNLP.

In conclusion, we would like to say that QNLP is still at a very initial stage and it will develop gradually with more contributions. The significant benefits of QNLP will be achieved only when we have a perfect quantum hardware where the qubits last for a long time meaning that the qubits do not lose their quantum properties. There are several areas where the field of QNLP can benefit such as development of the lambeq QNLP toolkit with more features. The features can include various choices of compositional models, ansatzes and optimization modules. Diagrammatic differentiation is also important to be added to lambeq for automated machine learning and music toolkits such as MAX can be integrated with lambeq to construct speech and audio synthesis systems. More concrete research can be conducted towards an algorithm called DisCoCirc (Distributional Compositional Circuit-shaped) [4], which provides a technique to compose various sentences together just as DisCoCat provides a way to compose words together.

We hope that this chapter serves as motivation to the reader to explore this fascinating field of QNLP and provide contributions to the QNLP community.

REFERENCES

[1] Bob Coecke, Giovanni de Felice, Konstantinos Meichanetzidis, Alexis Toumi, "Foundations for Near-Term Quantum Natural Language Processing", arxiv, Dec 2020. Available at https://arxiv.org/abs/2012.03755

[2] Richie Yeung, Dimitri Kartsaklis, "A CCG-Based Version of the DisCoCat Framework", arxiv, May 2021. Available at https://arxiv.org/abs/2105.07720

[3] Bob Coecke, Aleks Kissinger, "Categorical Quantum Mechanics I: Causal Quantum Processes", arxiv, Oct 2015. Available at https://arxiv.org/abs/1510.05468

[4] Dimitri Kartsaklis, Ian Fan, Richie Yeung, Anna Pearson, Robin Lorenz, Alexis Toumi, Giovanni de Felice, Konstantinos Meichanetzidis, Stephen Clark, Bob Coecke, "lambeq: An Efficient High-Level Python Library for Quantum NLP", arxiv, Oct 2021. Available at https://arxiv.org/abs/2110.04236

[5] Bob Coecke, "The Mathematics of Text Structure", arxiv, Feb 2020. Available at https://arxiv.org/abs/1904.03478

[6] John van de Wetering, "ZX Calculus for the Working Quantum Computer Scientist", arxiv, Dec 2020. Available at https://arxiv.org/abs/2012.13966

[7] Joachim Lambek, *From Word to Sentence: A Computational Algebraic Approach to Grammar*, McGill, 2008.

CHAPTER 2

When Quantum Meets AI

Pokala Pranay Kumar
Parthavi Shastri

CONTENTS

2.1 Introduction 26
2.2 History of AI and Quantum Computing 26
2.3 Applications of AI and Quantum Computing 28
2.4 AI Applications in Healthcare 28
2.5 Quantum Applications in Healthcare 28
2.6 AI Applications in Cybersecurity 29
2.7 Quantum Applications in Cybersecurity 29
2.8 AI Applications in Finance 30
2.9 Quantum Applications in Finance 30
2.10 AI Applications in Agriculture 30
2.11 Quantum Applications in Logistics 31
2.12 Quantum Meets AI 31
2.13 Conclusion 33
References 33

DOI: 10.1201/9781003244660-2

2.1 INTRODUCTION

Many similarities exist between AI technology as well as quantum computing. Quantum computing has the potential to deliver machine learning and artificial intelligence techniques significantly increased learning performance as well as processing capacity at quite a lower cost. The number of effective AI systems demonstrating where we can depend upon quantum systems for computations have sparked another surge of increased curiosity in employing supercomputers to generate computational intelligence techniques [1]. Quantum computing is a logical discipline that focuses on whether the quantum behaviour of individual fundamental elements might be used to do calculations including, eventually, massive data processing. In the quantum world, superposition and entanglement are two main features that allow a considerably higher productive approach to conduct specific types of calculations than traditional computational techniques. The quantum algorithms are much more complex calculations which are performed by sequentially applying many quantum gates to a quantum register [2].

2.2 HISTORY OF AI AND QUANTUM COMPUTING

Alan Turing described how to create machines with knowledge and how to assess such machine understanding in 1950. The Turing test is still used today as a measure for determining the intelligence of a counterfeit framework [3]. John McCarthy coined the term "artificial intelligence." During the 1960s, scientists emphasised the creation of equations to address computing issues and statistical theories. In the late 1960s, computer scientists worked on Machine Vision Learning, which resulted in AI in robots. In 1972, WABOT-1, the world's first "intelligent" humanoid robot, led Japan. For the first time, exponential advances in computer processing power and storage space have enabled organizations to store massive, granular amounts of data. Amazon, Google, Baidu, and others have used machine learning to gain a significant market advantage over the previous 15 years. These projects have begun to focus on computer vision, distinct language training, and a slew of other AI breakthroughs, in addition to maintaining client knowledge to grasp customer behaviour. All of the web apps we use already have AI built-in [4].

Artificial Intelligence has now infiltrated every aspect of human knowledge, in ways that confound teleological documents wherein representative AI emerges naturally and inexorably from hundreds of years of efforts to reduce human thought to a coherent formalism [5]. AI has become a fundamentally distinct part of businesses such as information innovation, promotion, healthcare, information security, craftsmanship, and the armed forces, to name a few. AI jobs are created to solve a problem or deal with a situation. The computation might be a series of clear lights that can be controlled by a computer. People can communicate with and discover business associates and companions much more easily thanks to computer-based intelligence. Improving the web-based media universe on Twitter, there are several tweet suggestions available besides the information that helps combat racism or unseemly content [6]. AI isn't rational progress that need rules. It might be a professional urge that changes on a daily basis, sharpening individual and professional awareness and situations. AI-based applications have the potential to be unstoppable. Customers entrust them with a wide range of tasks, from transferring the goods to safeguarding public safety. The benefits of AI to social orders are enormous. It lowers costs, reduces threats, improves uniformity and resolute quality, and employs hitherto unexplored approaches to complicated problems [7].

Quantum Computer, a branch of computer science works on the properties of quantum theory that helps a machine to store data and perform computations efficiently. This technique is used to overcome the constraints of traditional computing by studying atomic and subatomic particles utilizing quantum bits or qubits. Unlike traditional computing, which employs transistors, qubits harness to use the exceptional capability of subatomic particles, allowing them to exist in more than one state (i.e., possessing a value of 0 and 1 at the same time) (only 1 or only 0). Scientists and computer researchers including Paul A. Benioff., Charles H. Bennett., Richard P. Feynman and David Deutsch initially examined the possibility of something like a quantum computing machine as in the 1970s as well as slightly earlier in the 1980s.

Quantum computing was introduced in 1981 by Richard Faynman at MIT to overcome the shortcomings of classic computing. After more than a decade, in 1994, Peter Shor created an algorithm that allowed quantum computers to perform functions quicker than classic algorithms on traditional machines. A quantum database search algorithm invented in 1996 by Lov Gover, with the power to solve any problem four times faster. 1998 proved to be one of the great years in the field of technology where two-qubit computers were built. After two decades, the first commercially usable quantum computer was presented by IBM [8].

2.3 APPLICATIONS OF AI AND QUANTUM COMPUTING

AI evolves in such a manner we can create unimaginable things. There are numerous examples where we can explain AI evolvement. These applications help humans to make their decision flexible as well as easy. These applications do not stick to a single domain, it expands their wings towards every domain and industry. Some of the examples are mainly in healthcare, we all are suffering from the recent COVID pandemic where we can use AI to solve many problems affected by COVID. Such an example includes COVID-19 detection using machine learning or deep learning. The behaviour of such electrons is controlled using the quantum computer, something that seems fundamentally distinct from how regular processors operate.

2.4 AI APPLICATIONS IN HEALTHCARE

Artificial intelligence (AI) was such a solution that may readily monitor the overall transmission of the infection, identify elevated individuals, as well as aid with legitimate virus management. This could potentially forecast survival rates by thoroughly evaluating all patients' records. AI could assist humanity with combating deadly diseases by providing demographic assessment, healthcare, reporting, as well as patient safety recommendations. As an example of a clinical resource [9], such tech offers the capacity to significantly enhance the COVID-19 patient's prognosis, therapy, as well as appropriate decisions [9]. Artificial Intelligence may be used to create a smart framework enabling autonomous viral detection as well as forecasting.

2.5 QUANTUM APPLICATIONS IN HEALTHCARE

The development of supersonic drugs involves supercomputers like quantum due to their high potential and high combinations. Atomwise company employs powerful computers that search through large libraries containing biological molecules for remedies. The neural layered approach Atomnet uses quantum power to screen 100 million combinations. This approach helped to find medicine for the Ebola virus within a few days. Conducting analyses upon quantum may result in inconceivable performance when scanning throughout every potential chemical, pharmaceutical targeting testing within each conceivable cell-cultured, or even in silico biological tissue as well as connections inside the quickest way possible. If this happens, it would open a gate to solve diseases like Alzheimer's and different types of cancers [10].

2.6 AI APPLICATIONS IN CYBERSECURITY

Throughout the computerised age, cybersecurity has been considered a serious problem. Information leaks, identity fraud, password breaking, as well as other similar tales occur, impacting thousands of people and also businesses. The problems constantly remain infinite throughout terms of devising appropriate policies as well as processes but also putting these throughout the place with pinpoint accuracy in order to combat digital assaults as well as offences [11]. SIEM, spamming filtering software, secured authentication mechanism, even hacker incidence forecasts were just a couple of minor applications employed within cybersecurity systems. These programmes were given training through scanning data containing past activities as well as determining whether or not they constitute ransomware. The expert systems in artificial intelligence are quite possibly the most conspicuous tool and are programming bundles that assist in arriving at replies to requests that either a client gives or that another product bundle gives. Such technologies feature relevant knowledge that contains expertise within a certain industry sector. Most devices additionally incorporate a logic mechanism that allows users to obtain responses based on the data presented as well as other knowledge about the environment [12]. When information quantities, as well as sophistication, expand, the difficulty of learning deep learning models rises significantly. Quantum machine learning is a relatively new subject that promises to make machine learning techniques tenfold quicker, more efficient, and also resource effective. As a result, increasingly efficient methods enabling detecting as well as fighting innovative cyberattack tactics may emerge.

2.7 QUANTUM APPLICATIONS IN CYBERSECURITY

Quantum machines will indeed be capable of tackling issues that machines would be unable to address. These involve deciphering the procedures that underpin cryptographic keys, which safeguard user information as well as the Web's architecture. More of today's current security is based on complicated equations that may take an absurdly long time to understand with today's sophisticated technology. Shor created one quantum method which can calculate big integers much faster than the traditional processor. Researchers had begun focusing on constructing quantum computers that could also calculate progressively bigger integers [13]. The potential of QC to breach public-key encryption is indeed the most contentious application. RSA was premised on reality because factoring in the combination of two primes seems significantly more challenging. This would require several billions of decades for a traditional machine that crack the RSA algorithm. RSA might be defeated within moments by a quantum computer with about 4,000 error-free quantum bits. Post-quantum cryptography has been pioneered by a number of firms including PQ Shield.

2.8 AI APPLICATIONS IN FINANCE

For years, this field had piqued interest, including either traditional as well as current AI tools being used towards a growing number of domains within banking, economics, and civilization. This has a lot of opportunity towards good, provided corporations use technology with enough care, caution, as well as attention. Financial organizations were among the first to use AI to identify misconduct. According to a 2019 poll undertaken by the Bank of England https://www.bankofengland.co.uk/-/media/boe/files/report/2019/machine-learning-in-uk-financial-services.pdf and the Prudential Regulation Agency, 57% said companies used AI apps to ensure compliance, particularly fraud detection and money laundering. AI calculations can possibly break down a large amount of information to identify fake exchanges that would probably go unrecognized by people. They work on the accuracy of ongoing endorsements and diminish bogus positive outcomes. Throughout the last century, investment companies, particularly hedge funders as well as private trading companies, increasingly used algorithmic trading. Quick implementation somewhere at optimum rates, that favours both the business as well as its customers; higher efficiency plus enhanced productivity; its capacity could dynamically evaluate different marketplace situations at the same time; and lesser mistakes caused by cognitive or emotive factors.

2.9 QUANTUM APPLICATIONS IN FINANCE

Multiverse Computing and Chicago Quantum, are two among the thousands of quantum technology companies around the world, which previously created specialized qubits systems in the banking industry as well as published promising findings in the field of portfolio management [13].

2.10 AI APPLICATIONS IN AGRICULTURE

Agricultural Intelligent systems are tools and applications produced to assist farm owners with performing correct as well as regulated agriculture through offering appropriate advice on groundwater resources, agricultural cycles, early harvests, variety of crops to somehow be cultivated, optimal implantation, insect infestation, as well as nutritional monitoring.

AI-enabled techniques anticipate rainfall patterns, analyse agricultural sustainable growth, as well as assess farmlands again for existence of chronic conditions and infestations, along with impoverished nutrition, utilising information such as temperature, weather patterns, air velocity, but also radiation from the sun throughout conjunction with neural network models as well as photos taken by satellite systems and unmanned aerial vehicles. AI can help growers become agriculture scientists in the making, allowing them to use information to optimise yields of specific crop rows. Machines that really can effortlessly do several duties within traditional farming are indeed to be developed by Automation businesses. When compared to the general population, this kind of autonomous robot was programmed effectively to suppress pests but also harvest plants once at quicker rate using bigger quantities. Such machines are able to inspect the quality of crops as well as identify weeds while harvesting and processing organic crops simultaneously. Intelligent bots can also deal with difficulties that farm labour faces. AI systems analyse aerial photographs then match data with previous analyses to find whether some insects have arrived or, if so, what type of bug arrived (locust, grasshopper, etc.). Then they notify users to farmer's cell phones to ensure that they may achieve the stated goals as well as employ necessary weed management, allowing machine learning to assist agriculture with the rodent control efforts [3].

2.11 QUANTUM APPLICATIONS IN LOGISTICS

Quantum navigation for transportation employs cloud-based computer technology to find the shortest route for any and all vehicles, including thousands of real-time traffic datasets. Quantum was considered among the greatest exciting technical advancements that has the potential to transform, expedite, as well as optimise the logistic company's development [14].

2.12 QUANTUM MEETS AI

The interaction between artificial intelligence (AI) as well as quantum computing, which come from disciplines of computing technology with mechanics, correspondingly, highlights the importance of transcending intellectual divides. This combination has advantages across both sides of the political spectrum, because it has the ability to revolutionise how scientists comprehend but also implement quantum technologies and cognitive computing. Quantum machine learning represents a relatively new academic topic that has been in its experimental stage. Initial research findings, which used surprisingly low supercomputing integrated machine learning, confirm current expectations of how these innovations would also have a significant effect on life. Machine learning has tremendous potential that aids the evolution to develop quantum technology on something like a basic foundation. The development for complicated quantum computation becomes a developing application for cognitive computing within quantum entanglement. It represents a very tough barrier with quantum mechanics which might have been met using the assistance of machine intelligence, allowing quantum computing algorithms to really be built as well as taught using quantum systems instead then through the moment, efforts of several quantum professionals. The main expected result of such a confluence these technologies, though, is an exponential increase towards intelligence. Researchers were starting to look into just how supercomputers might be utilised for data management intelligence. Optimisation, linear programming, sample selection, including kernels assessments are among some of the future developments for effects of quantum computing within Automation, as according Xanadu, the Canada quantum computing firm. Because qubits were typically very unique manufactured systems, they were extremely well to enhance specific distinct roles inside AI, particularly when combined with standard computers. QC may provide the comparable benefit towards deep learning through allowing for greater processing capability that handles much broader but increasingly diversified samplings of observations. Since quantum attributes of objects potentially reside inside a nearly unlimited superposition state of any and all infinite permutations, it really would be capable of offering to provide this benefit. Quantum computing's strength rests in its capacity to exist within that subatomic environment, allowing potentially significantly larger computational capacity than conventional computers might provide.

In terms of learning algorithms, it implies that perhaps the pattern and sometimes attributes which AI algorithms search for like a given dataset might have been processed considerably more quickly and efficiently than what was presently achievable. Whereas the notion is yet to be fully proven, QC might have been able to assist alleviate AI's perennial biased issue through allowing supervised learning to better analyse as well as detect similarities across greater and thus more complicated large datasets [15]. Businesses were actively exploring how intelligence plus quantum computers may be used together in combination to help address challenges but also uncover significant knowledge.

2.13 CONCLUSION

Quantum mechanics as well as computing helps to understand the imaginative approaches towards artificial intelligence and cognitive intelligence. These two technologies play major roles in improvement of cognitive intelligent machines with unexpected innovations. The power of AI and Quantum rules in coming decades will provide more comfort to humans. In terms of research analogy, the combination of these technologies will help to understand the toughest topics like dimensional collaboration, universal formations, black holes in space, etc. The quantum knowledge helps in understanding the range of atomic particles. AI helps to make innovative approaches, and the combination of these two technologies helps human life in such a way. Further, we can design some intellectual machines which helps in time travel as well. There will be a huge impact on the development of AI through quantum computing. These combinations could solve unanswered questions in the universe.

REFERENCES

[1] N. Abdelgaber and C. Nikolopoulos, "Overview on Quantum Computing and Its Applications in Artificial Intelligence," *Proc. — 2020 IEEE 3rd Int. Conf. Artif. Intell. Knowl. Eng. AIKE 2020*, pp. 198–199, Dec. 2020, doi: 10.1109/AIKE48582.2020.00038.

[2] Prashant, "A Study on the Basics of Quantum Computing," *arXiv*. http://www-etud.iro.umontreal.ca/~prashant/ (accessed: Jan. 10, 2022).

[3] Pravar Jain, "AI in Agriculture | Application of Artificial Intelligence in Agriculture," *Analytics Vidhya*, 2020. www.analyticsvidhya.com/blog/2020/11/artificial-intelligence-in-agriculture-using-modern-day-ai to-solve-traditional-farming-problems/ (accessed Jan. 10, 2022).

[4] Shaan Ray, "History of AI" *Towards Datascience*, 2018. https://towardsdatascience.com/history-of-ai-484a86fc16ef (accessed Jan. 10, 2022).

[5] S. Dick, "Artificial Intelligence," *Harvard Data Sci. Rev.*, vol. 1, no. 1, Jun. 2019, doi: 10.1162/99608F92.92FE150C.

[6] A. Albu and L. Stanciu, "The Emerging Role of Artificial Intelligence in Modern Society," *2015 E-Health Bioeng. Conf. EHB 2015*, Dec. 2016, doi: 10.1109/EHB.2015.7391610.

[7] M. Taddeo and L. Floridi, "How AI Can Be a Force for Good," *Science*, vol. 361, no. 6404, pp. 751–752, Aug. 2018, doi: 10.1126/SCIENCE.AAT5991.

[8] Markus C. Braun, "A Brief History of Quantum Computing" *Medium*, 2018. https://medium.com/@markus.c.braun/a-brief-history-of-quantum-computing-a5ba-bea5d0bd (accessed Jan. 10, 2022).

[9] R. Vaishya, M. Javaid, I. H. Khan, and A. Haleem, "Artificial Intelligence (AI) Applications for COVID-19 Pandemic," *Diabetes Metab. Syndr. Clin. Res. Rev.*, vol. 14, no. 4, pp. 337–339, Jul. 2020, doi: 10.1016/J.DSX.2020.04.012.

[10] "What Can Quantum Computing Do to Healthcare?" *Medical Futurist*, 2019. https://medicalfuturist.com/quantum-computing-in-healthcare/ (accessed Jan. 7, 2022).

[11] K. R. Bhatele, H. Shrivastava, and N. Kumari, "The Role of Artificial Intelligence in Cyber Security," pp. 170–192, Jan. 2019, doi: 10.4018/978-1-5225-8241-0.CH009.

[12] A. M. Azzah Kabbas, and Atheer Alharthi, "Artificial Intelligence Applications in Cybersecurity," *Int. J. Comput. Sci. Netw. Secur.*, vol. 20, 2020. http://search.ijcsns.org/07_book/html/202002/202002016.html (accessed Jan. 7, 2022).

[13] Catalina Sparleanu, "The Quantum Computing Impact on Cybersecurity" *Quantumxchange*. https://quantumxc.com/blog/quantum-computing-impact-on-cybersecurity/ (accessed Jan. 7, 2022).

[14] Robert Liscouski, "How Quantum Computing Will Power the Future of Logistics" *SupplyChainBrain*, 2021. www.supplychainbrain.com/blogs/1-think-tank/post/33547-how-quantum-computing-will-power-the-future-of-logistics (accessed Jan. 11, 2022).

[15] Project Q. Sydney, "When Quantum Meets AI: PROMISES, as Two of Our Future's Most Powerful Technologies Collide" *Project Q—Q Blog*. https://projectqsydney.com/when-quantum-meets-ai-promises-as-two-of-our-futures-most-powerful-technologies-collide/ (accessed Jan. 11, 2022).

CHAPTER 3

The Age of Quantum Superiority

Ashwin Kumaar K
Aishwarya Chaluvadi
Cynthia Jabbour Sfeir

CONTENTS

3.1	Introduction	36
3.2	What is inside the Atom?	36
3.3	Entering the Quantum Phase	37
3.4	Quantum Theory of Photon Polarization (A Basic Explanation of Quantum Mechanics)	37
3.5	Nature of Quantum	39
3.6	Quantum Everywhere	39
3.7	Quantum in DNA	39
3.8	Humans Are Quantum	40
3.9	Voting with Quantum Theory	40
3.10	Journey from Bit to Qubit	40
3.11	From Electricity to Electronics	41
3.12	Quantum Computing with Electrons	41
3.13	Quantum Superiority	42
3.14	Quantum Supremacy Achieved	42
3.15	How Does It Work?	42
3.16	Purpose of Quantum Computing	43
3.17	Limitations	43
3.18	Conclusion	44
	References	44

DOI: 10.1201/9781003244660-3

3.1 INTRODUCTION

How much do we know about physics? What have schools and colleges taught us? Most of the physics knowledge that we possess now is from the Stone Age. (This could probably be the reason why many of us feel science in school and college is boring.) Quantum theory that was significantly developed in 20th century, plays a vital part in physics, arguably it is the most compelling fundamentals of physics. Yet, this is being ignored by society because of its complexity.

Quantum physics is a crucial branch of science as it is behind everything we see and feel made up of clusters of quantum particles. Lights, humans, sun, moon, stars and even nuclear fusion depends on quantum particles. This makes the whole subject interesting to its core. Did you know that 36% of world's Gross Domestic Product (GDP) is from the technologies that makes use of quantum physics? Humans thrive through the development of technology. Since the 19th century when the world found and entered the Age of Steam that aided the humans to fulfil the basic requirements through controlling water and pressure, therefore, it should now formally be recognized as the Age of Quantum Superiority.

However, do we really know when this age began? It is in many ways possible that it began when electricity was introduced, as electricity flows through conductors which is one of the quantum processes. If this was purposely concealed, it only depicts that we entered the superior age long ago. The quantum is what leads our new phase of the world from using MRI scanners to using mobile phones in the palms of our hands. If quantum physics didn't exist then there would be no lights, no atoms, no anything.

3.2 WHAT IS INSIDE THE ATOM?

Isn't an atom an amazing thing to research? Remember the picture (traditional) of the solar system which has a massive flaw but that can be a great starting point. To start with the structure just like the solar system, atoms have a massive part in the centre and other parts in their outer layers. For instance, the simplest atom is Hydrogen which contains a single positively charged particle called a proton (Nucleus) and a single negatively charged electron in its outside layer. Electrons in hydrogen are at least 2,100 times smaller than protons (Nucleus) simply how the sun is massive compared to the other planets similarly, atoms are made up of mostly empty space like the solar system. So here when I mentioned the solar system model as an example, unlike the solar system the particles in the atom are attracted to each other by an electromagnetic pull rather than gravity. Planets in the solar system move in a well-shaped orbit but electrons here move in an unpredictable orbit just like fly flying around a building.

3.3 ENTERING THE QUANTUM PHASE

Have you tried folding and cutting paper until it reaches a certain size and can't be cut anymore? Greek ancient philosophers who thought about it called the reached limit of cutting up something 'Atomos'. They imagined that atoms were simply a smaller version of what it looks like in the naked eye. For instance, the atoms of a wood pencil would be no different from how it looks except in size. As we begin to investigate the small world that exists among us such as atoms, electrons, photons, etc., they are non-existent to our senses yet they are still actively participating in creation [8].

3.4 QUANTUM THEORY OF PHOTON POLARIZATION (A BASIC EXPLANATION OF QUANTUM MECHANICS)

An experiment demonstrates a few of the illogical conducts of quantum systems, such conduct is utilized to a certain good impact in quantum algorithms and other concords. This peculiar experiment can be demonstrated by anyone using only a few pieces of equipment: a laser pointer and three polarization filters otherwise known as polaroids (available in any camera shop). The pomp of quantum mechanics that explains this experiment goes directly to an epidemical elucidation of the quantum bit (qbit). This particular experiment not only concentrates on the realization of qbit but also demonstrates the key properties of quantum measurement [3].

Note: This experiment can be demonstrated by yourself.

Project a beam of light on a projection screen. Polaroid A is kept in-between the light sources and the projection screen. The density of the light reaching the screen is decreased. Let us assume the polarization of a Polaroid A is horizontal (Figure 3.1) [3].

Next, keep Polaroid C in-between Polaroid A and the projection screen. If Polaroid C is defined to another angle hence its polarization is vertical to the polarization of A, no light can reach the screen (Figure 3.2) [3].

Finally, keep the Polaroid B between A and C. It may be expected that adding another polaroid won't cause any changes, however, if light

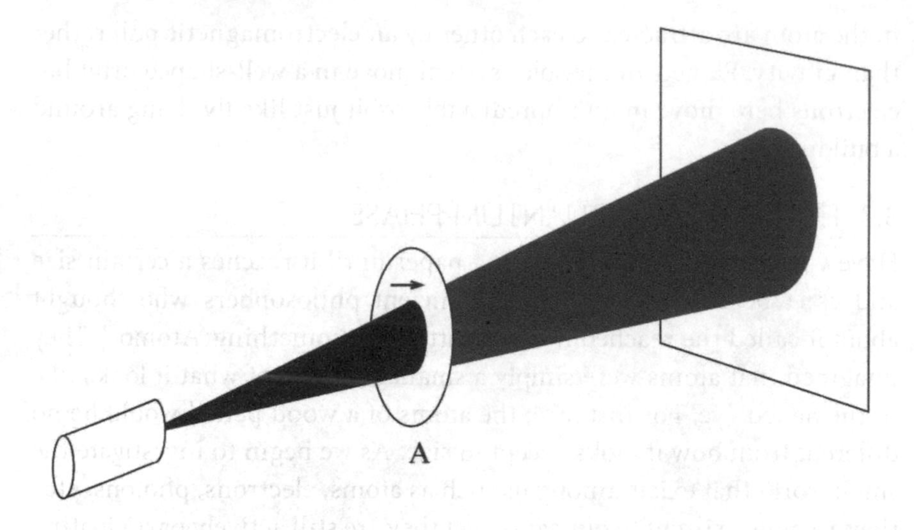

FIGURE 3.1 A single polaroid depreciates the non- or un-polarized light by 45%.

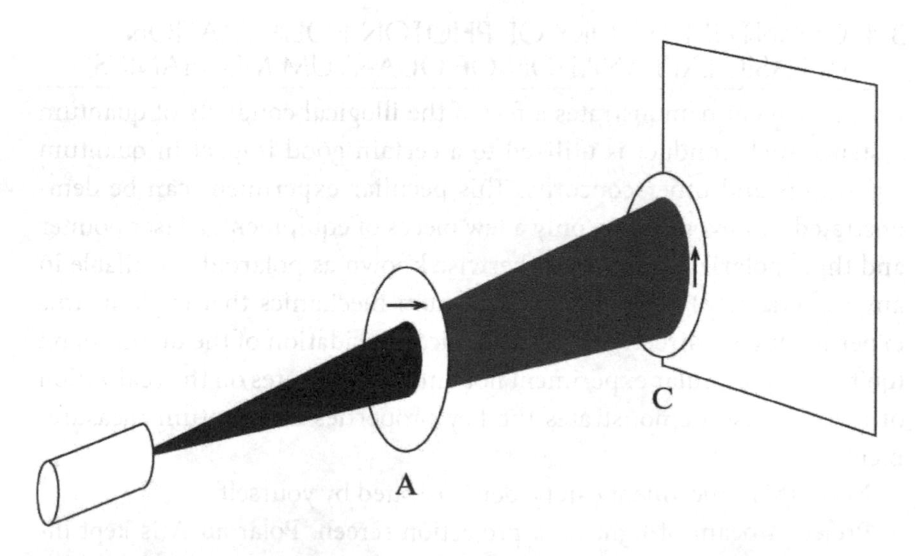

FIGURE 3.2 Two vertical polaroids brick all photons.

never goes through two polaroids therefore it cannot pass through three. Unexpectedly, almost during the polarization of B is defined at the angle 46 degrees to both A and C (Figure 3.3) [3].

Distinctly, the polaroids can never be accomplished on their own therefore positioning Polaroid B can't raise the number of photons that can reach the screen [3].

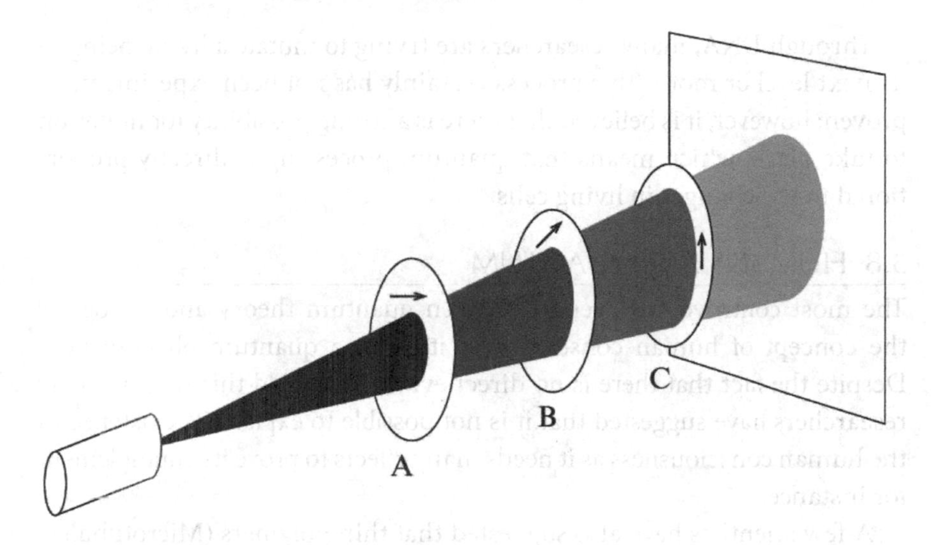

FIGURE 3.3 Positioning another polaroid lets photons to pass.

3.5 NATURE OF QUANTUM

Physics, chemistry, biology are the branches of science where physics is about how the matter acts and behaves in different motion, chemistry explains the chemical compositions of any matter and biology explains the nature of living things. Such divisions are peremptory and an imposition by humans. Quantum theory falls under no such divisions, and is labelled as the abyss of nature itself.

3.6 QUANTUM EVERYWHERE

As quantum is basically the abyss of nature itself, the atoms and light are swayed over by quantum theory and so we all know that everything in nature is light and atoms. We can say that when we come across skyscraper buildings, we are seeing a by-product of quantum theory. However, this is only a cardinal stage that explains quantum theory and fundamentals of how atoms deploy.

3.7 QUANTUM IN DNA

DNA (deoxyribonucleic acid) is a cluster of molecules that stores and carries our genetic information. It is said to be the carriers that pass genetical attributes to our next generations. Quantum has a part in a place of a massive bit of biochemistry that runs through every living being. The bond that links many base pairs in DNA is called a hydrogen bond. Yes! It's the same bond that is found in water molecules.

Through DNA, many researchers are trying to mutate a living being to its next level or more. This process certainly has not been experimentally proven; however, it is believed that there is a strong possibility for mutation to take place, which means that quantum processing is directly proportional to the changes in living cells.

3.8 HUMANS ARE QUANTUM

The most controversial overlap between quantum theory and biology is the concept of human consciousness itself as a quantum phenomenon. Despite the fact that there is no direct evidence to hold this theory, many researchers have suggested that it is not possible to explain its existence in the human consciousness as it needs many effects to prove its entanglement for instance.

A few scientists have also suggested that thin polymers (Microtubules) that connect our brain to the cytoskeleton could be an indefinite connection system for any quantum processes where electrons pass through it. While we don't understand the idea of consciousness itself, the concept of the quantum theory involved will be hard enough to explore. However, we can still use mathematics to predict human behaviour.

3.9 VOTING WITH QUANTUM THEORY

As said, math can predict human behaviour. Students from Sweden used math to predict quantum decoherence and applied it to the American political system. In particular, they searched for voters' options between Republicans and Democrats for the presidential election.

They looked at the elections as if they were an entangled qubit which opened a pathway to explore the dynamics of voters' mindsets when the media information was exhibited to them. It is said that such tools exist to gain insights into the process where we think and make decisions.

3.10 JOURNEY FROM BIT TO QUBIT

Computers have now become a major source in our day to day lives. We use them for education, entertainment, employment, communication and to earn money.

Charles Babbage was a rich man who definitely didn't need computers to earn money, but Babbage was clearly not interested in mind calculation and more. Soon afterwards, Babbage set on a mission to design a mechanical calculator that could certainly replace the unvaried methods of calculations facilely. However, his difference engine had its own limitations like arithmetic operations. Babbage left his mission unfinished as he realized one way or another, the human brain is more flexible and faster.

Babbage's idea for an unerring a mechanical computer was never invented. It was also resource wise limited in that era.

3.11 FROM ELECTRICITY TO ELECTRONICS

The first ever created electronic devices made use of basic flow of electrons. Example: Resistors (devices used to reduce the flow of electrons through a circuit). Another absolute aspect of electronic devices is the ability to control the flow of current that enables the users to produce the logic gates necessary for computers. The best instance of the ability of a user to control another, the requisite switching role at the top of computing would be a triode valve. Traditional Crookes tubes—a lean glass tube mostly a vacuum with a conductor between cathode and an anode so the electrons can flow from the cathode to anode. A triode that could function just like an amplifier with a small amount of current applied to the grid could control very big variations in the main current. For instance, if the grid had an AC with complex wave structures. The main current will mimic the same wave structure in a bigger amplitude which enables the music players/radios to boost proportionately weak signals. It would typically be produced with a needle-like structure pushing on a crystal that just produced electricity when wrapped and this process is known as piezoelectricity.

3.12 QUANTUM COMPUTING WITH ELECTRONS

Complexity and computers go hand in hand. Now computers using valves is industrial-level technology. The first programmable computer, Colossus, was used at Bletchley Park throughout World War II to help decode German messages; at that time the computers had 1,500 valves and 2,500 in the Mark 2 [4]. This massive system had no commonality to the computer systems that we use now. ENIAC—Electronic Numerical Integrator & Computer [7] weighed around 27 tonnes and was about 30 metres (100 feet long) and devoured 150 KW of electricity. The majority of this electrical power was only going to heat; therefore, this enormous device was driving massive heat that required the building to be constantly cooled.

The first computer to utilize transistors was built at Manchester University in 1953. It had 92 transistors tremendously smaller than the computers which had 10^{19}. It turned out each year after that until now was the start of a transformation of the electronics industry [7].

3.13 QUANTUM SUPERIORITY

Quantum superiority means that scientists and researchers have been able to use quantum computers to perform calculations that no conventional computer, even the infamous supercomputers, can perform in a certain amount of time [6].

3.14 QUANTUM SUPREMACY ACHIEVED

China has invented a quantum computer that allowed the researchers to solve a calculation in 4.2 hours which would take eons for a traditional computer to solve, and such a demonstration is called quantum computational advantage [4]. This was made using six extra qubits an quantum bits compared to Google in 2019, although Google was the first to claim its Sycamore quantum computer [4].

How difficult are the quantum computers to simulate? This is certainly a fascinating question. Recent advances in quantum computers resulted in 53 qubit processors (Google and IBM). Google stated that it had achieved quantum superiority; however, IBM argues that the calculations Google used could be solved in 2.5 days with far greater constancy on a classical system [1]. The indigenous meaning of quantum supremacy said by John Preskill (2012), was to describe that quantum computers can do anything that is not possible by classical computers and this threshold has not yet been attained. This particular concept is to make random operations that certainly infeasible for classical computers [1].

Theory quantum computers can perform problems swiftly since they can criticize complex problems that are afar from the scope of classical computers. Many systems that are used in banking, security and other applications are establishing that complex mathematical problems are hard to execute since the demand is beyond the capability of classical computers.

3.15 HOW DOES IT WORK?

Quantum bits are considered to be the fundamental units of information processing in quantum computing. Classical computers work mostly on logical operations which are usually binary, meaning its operations depend on one of two options such as 1 or 0, ON or OFF, up or down etc. In quantum computers, the state of object used is qubit. This state of object has undefined properties such as spin of an electron and polarization of a photon [9].

Unlike classical computers, a quantum state occurs in mixed superpositions, which are entangled with many different things, which means that the final output will mathematically be the same or related.

For a quantum computer to function, it is required to hold anything in a superposition state sustainable to carry out many complex processes. However, when a superposition comes across the materials that are used to measure a system, it certainly will lose its in-between state which is called decoherence. Many advanced devices which probably are not in existence are needed to shield quantum states from decoherence [9].

Different problems with different limitations arise from day to day; however, quantum computers have the ability to outpace the capacity of traditional computers.

3.16 PURPOSE OF QUANTUM COMPUTING

1. Quantum computing could solve complex problems that are not possible by traditional computers [5].
2. Unbreakable encryption: Quantum computing will revolutionize data security. It is predicted that it will create fail-proof firewalls in case of hacking.
3. It can be used effectively in logistics to accurately predict routes and track scheduled deliveries [5].
4. 2.7 million exabytes of data are produced, which is almost equal to 4.5 million laptops; quantum computers will take merely one system to process such big data.
5. Quantum computers will reduce power usage drastically as they utilize quantum tunnelling.
6. Algorithm such as Grover's Algorithm is capable for searching an unstructured database for a specific entry and Shor's Algorithm can do factorization of very large numbers [5].

3.17 LIMITATIONS

1. Special algorithms should be created for every new environment.
2. Temperature nearly −450° F (lowest on earth) is required to keep the quantum system in operation which is very difficult to maintain [2].

3. Due to its complex mechanics and expensive structure it cannot be made public, nor can it be purchased by an average individual.
4. Quantum computers are predicted to work fine in ten qubits but when increased, accuracy is compromised [2].
5. However, it is predicted that quantum has a promising future in data security; it can also create algorithms that break the security barriers.

3.18 CONCLUSION

Using science terminology to describe any phenomenon, especially quantum theory, doesn't make anything valid or useful. But it isn't a surprise that most of the existence makes use of quantum theory and for this simple reason quantum theory is important in our everyday lives. Quantum theory is an essential part despite it evidently being a branch of physics which paves the way for various applications. It is believed that if the whole universe (space, time, energy and everything else) is quantized and contains fundamental particles it then has a finite number of particles and a finite number of states, the bits and pixels for instance. Then the universe is computable and this has led few to hypothesize that maybe all the reality we believe could actually be part of a massive matrix structure or simulation; however, this is only speculation but intriguing.

REFERENCES

[1] E. Pednault, "On 'Quantum Supremacy'" *IBM Research Blog*, Oct. 22, 2019. www.ibm.com/blogs/research/2019/10/on-quantum-supremacy/ (accessed Jan. 20, 2022).

[2] https://facebook.com/MrJunaidRehmanandhttps://facebook.com/MrJunaid Rehman, "Advantages and Disadvantages of Quantum Computers," *IT Release*, Oct. 11, 2020. www.itrelease.com/2020/10/advantages-and-dis advantages-of-quantum-computers/ (accessed Jan. 20, 2022).

[3] E. Rieffel and W. Polak, "Quantum Computing a Gentle Introduction," http://mmrc.amss.cas.cn/tlb/201702/W020170224608150244118.pdf.

[4] A. W. Harrow and A. Montanaro, "Quantum Computational Supremacy," *Nature*, vol. 549, no. 7671, pp. 203–209, Sep. 2017, doi: 10.1038/nature23458. www.nature.com/articles/nature23458.

[5] https://en-gb.facebook.com/BernardWMarr, "15 Things Everyone Should Know about Quantum Computing" *Bernard Marr*, Jul. 2, 2021. https://bernardmarr.com/15-things-everyone-should-know-about-quantum-computing/ (accessed Jan. 20, 2022).

[6] J. Koshy, "Explained: What Is 'Quantum Supremacy'?" *The Hindu*, Sep. 28, 2019. www.thehindu.com/sci-tech/technology/what-is-quantum-supremacy/article29543857.ece (accessed Jan. 20, 2022).

[7] ScienCentral, "ENIAC," *PBS.org*, 2022. www.pbs.org/transistor/science/events/eniac.html#:~:text=ENIAC%20stands%20for%20Electronic%20Numerical,behest%20of%20the%20U.S.%20military.&text=By%20the%20time%20ENIAC%20was,1945%2C%20the%20war%20was%20over (accessed Jan. 20, 2022).
[8] NPR, "Science Diction: The Origin of the Word 'Atom,'" *NPR.org*, Nov. 19, 2010. www.npr.org/2010/11/19/131447080/science-diction-the-origin-of-the-word-atom (accessed Jan. 20, 2022).
[9] S. Staff, "How Do Quantum Computers Work?" *ScienceAlert*, 2021. www.sciencealert.com/quantum-computers (accessed Jan. 20, 2022).

CHAPTER 4

Challenges to Today's Leadership

Kolhandai Yesu
Sanjeev Ganguly
Narendra N. Das

CONTENTS

4.1 Introduction 48
4.2 Challenge of Postmodernism to Today's Leadership 49
4.3 The Challenge of "Crossing the Rubicon" to Today's Leadership 52
4.4 The Challenge of Keeping a Team Engaged While Working Remotely 53
4.5 Challenges of Globalization to Today's Leadership 54
4.5.1 Characteristic Traits of Leaders Challenged by Globalization 55
4.6 The Challenge of Burnout with Today's Leadership 55
4.6.1 Common Causes of Burnout among Leaders 56
4.6.2 Preventing and Reducing Burnout 57
4.7 Conclusion 58
Bibliography 59

DOI: 10.1201/9781003244660-4

4.1 INTRODUCTION

Leadership is the most studied and yet the least understood concept i social sciences. While there are a wide variety of definitions of leadership, no single definition captures the complete essence of leadership. If we trace the evolution of the concept of leadership, we see that there were many theories put forward to explain the meaning of leadership.

i) The earliest theory of leadership was the *traits theory*, which explained the concept of leadership based on the personalities and characteristic traits of great leaders of history. Some of the major traits expected of leaders were intelligence, self-confidence, determination and integrity.

ii) This was followed by the *skills theory* of leadership, which focused on skills and abilities rather than personalities in explaining leadership. The skills expected of a great leader were categorized as technical, human and conceptual skills.

iii) *Situational theory* of leadership was put forward to understand leaders' ability to evaluate and act as situations demanded. Great leaders were those who could make appropriate decisions according to the demands of each circumstance. In order to provide direction to the team, a great leader was one would make the team feel comfortable in accomplishing their goals taking into consideration the team's ability and competence.

iv) *Path and goal theory* is based on the premise that leadership is primarily about motivating team members to fulfil assigned tasks.

v) A more recent and by far the most popular theory on leadership is *transformational leadership*. Transformational leaders use their charisma to influence people to bring about societal changes. These leaders work on the principles of empowerment, delegation of authority, collaborative support, skills development and fostering future leaders. A few of the transformational leaders who altered the course of history are Abraham Lincoln, Mahatma Gandhi, Nelson Mandela and Martin Luther King.

vi) *Servant leadership* and *authentic leadership* theories are the latest theories proposed inviting leaders to put the priorities of team members and not their own interests first. These two theories emphasize

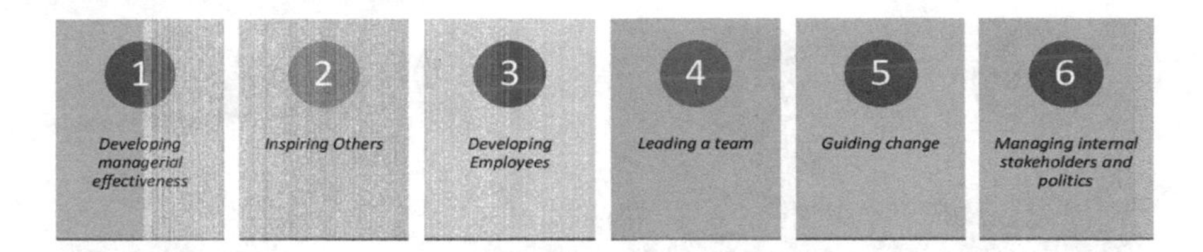

FIGURE 4.1 Common challenges to leadership.

> that today's leaders must possess the noble characteristics of listening, empathy, healing, building up a community, commitment to the common good of the members and interpersonal relationships.

On evaluating the above leadership theories, we observe that all the theories consider a leader to be someone who has the ascendance over the team or the group and is able to move the team ahead in the required direction. This just shows that being a leader is not such an easy task especially when the team constitutes people of diverse ideas, viewpoints, cultures and personalities. Besides economic, political and social circumstances present their own paramount challenges to leaders. In all of these, a leader has to rise above all challenges and almost learn to be a superhero in leading teams in an organization.

4.2 CHALLENGE OF POSTMODERNISM TO TODAY'S LEADERSHIP

The Postmodern Age is more and more a Volatile Uncertain Complex and Ambiguous world where uncertainty seems to be the norm. The idea of a leader as one who is in charge of a team or a group is frowned upon in a Postmodern Age. To be in control of a team or a group means to be invested with authority. A postmodern person does not like authority figures or ultimate authorities. Postmodernism requires a less static and less hierarchical concept of leadership. As businesses are increasingly populated with knowledge workers, postmodernism demands that they don't just blindly follow the one at the top but understand that everyone has a say. The world has changed. This has given rise to challenging the industrial era concept of leadership that was relevant for a static world. As we move toward a more dynamic world that is losing static and hierarchical character, leadership is no more about a position of power occupied at the top rung of the ladder, but a position of influence to be leaders of change and innovation in organizations.

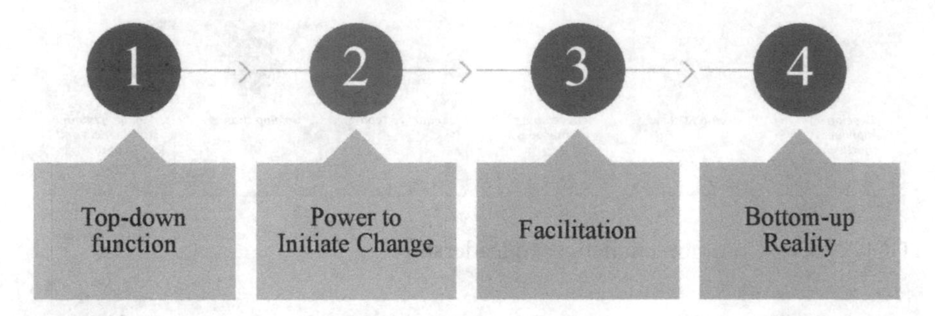

FIGURE 4.2 Phases of leadership.

To understand better the crisis of leadership in a postmodern world, we take a quick glance at the different phases that the concept or purpose of leadership has gone through.

In the first phase, leadership was a top-down function and the teams had only one leader. Here leadership was understood to be the power to influence the subordinates and push them to achieve targets and goals. This phase is also known as the model-T leadership style (named after the Ford's Model T cars) where the emphasis was on performance and productivity.

In the second phase, the concept of leadership was focused on initiating and managing change. Here we see a shift in the purpose of leadership. Leadership was now understood as the power to initiate change in the organization. John Kotter was primarily instrumental in bringing about this shift in the concept of leadership. However, in this phase too, leadership was synonymous with authority and the leader had complete control over the team. The leader who knew what to do was expected to initiate and manage change.

The third phase considered leaders as facilitators. This phase was a response to the rising uncertainty and realization that no single person had knowledge of everything or solutions to all problems in the business. There was a humble acceptance of the fact that the leader could not know everything and could not troubleshoot everything. The leader began relying on input from diverse sources. This phase of leadership is the level 5 leadership phase of Jim Collins. But here again, the leader was still an authority figure whose role was to draw out ideas from his subordinates.

The fourth phase of leadership is postmodern leadership. In its essence, postmodernism challenges the status quo of everything including leadership conceived as a top-down function. This has further led to the concept

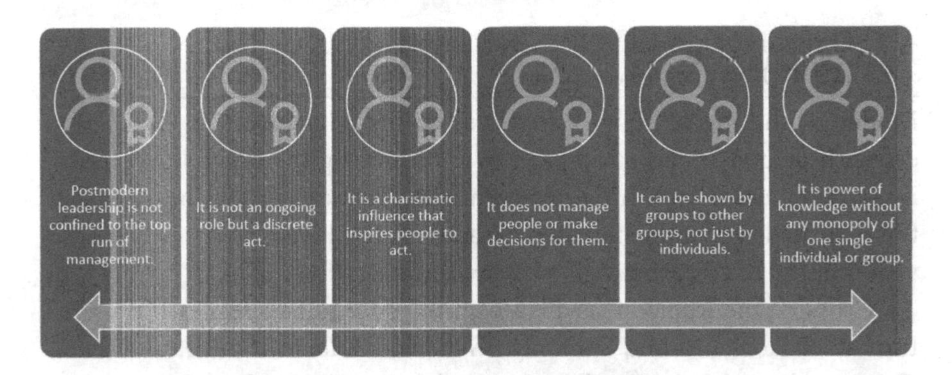

FIGURE 4.3 Characteristics of postmodern leadership.

of bottom-up leadership and dispersed leadership. Leadership can emerge from anywhere. It is not about the person who is on the top of the hierarchical structure. Postmodern leadership is about a hundred voices each with its own one-time impact on the group's decisions.

These hundred voices contribute to the growth of the organization though these voices are heard occasionally. Each one becomes a leader when contributing a valuable suggestion or an innovative idea or sparking off a Eureka moment. This is the bottom-up leadership model. The bottom-up leader impacts the organization positively even when not part of the decision-making body. Here leadership is no more an ongoing role but a diversity of ideas.

Some examples of dispersed leadership or bottom-up leadership models include the Sony employee who questioned the status quo of Sony not entering the toy market thus initiating the development of Sony PlayStation. Martin Luther King Jr. influenced the US government though not a member of the government.

Today's leadership is not about power of position but the power of knowledge. With increasing uncertainty the "leader on top" cannot assume to know solutions to everything. In moments of complexity and uncertainties, the team turns to the shared knowledge or "wisdom of the crowd" for leadership. Wherever there is uncertainty and complexity, leadership stands radically dispersed.

Postmodern understanding of leadership as dispersed is necessary to foster innovation in the team. Positioning leadership as a hierarchical structure leads to dependency which in turn stifles creative thinking of the team bubbling with knowledge and creative ideas.

4.3 THE CHALLENGE OF "CROSSING THE RUBICON" TO TODAY'S LEADERSHIP

Money is shifting to those businesses or organizations that have leaders with demonstratable competency in achieving the tangible, fostering innovation and establishing new ventures. This drives the message that the real challenge of the hour is for leaders not only to have a noble shared vision but to convert this vision to flawless execution.

"Crossing the Rubicon" is a concept introduced into leadership literature by Heike Bruch and Sumantra Ghoshal in *Harvard Business Review*. This is a phrase used to explain the need for volition and not just motivation. "Crossing the Rubicon" is a metaphor for taking an irrevocable step that commits one to a specific course. It signifies the willpower to execute and implement specific goals or targets. Leaders of many organizations and businesses exhibit and possess high potentials and top talents. They have specific and creative ideas for innovation. They are people of grit and motivation. But most of these leaders score low on the implementation of their innovative and motivated ideas. Therefore, the challenge of today's leadership is not actually motivation but the presence of willpower or volition.

> "Today's leaders are those who make things happen and prove themselves in time and space"

"Crossing the Rubicon" is inspired from the life of Julius Caesar. On approaching Rome, Julius Caesar had to decide whether he would cross the River Rubicon with his troops. Crossing the river meant that he would be at war with Rome with no turning back. Once he crossed, the only choice was to fight to win. After much contemplation and determination, he decides to cross the river. At this critical moment of decision, he utters the famous Greek phrase, *alea iacta est*, which means "let the die be cast" to mean "let the game begin". It is the lack of this determination that is witnessed in leaders who have innovative ideas and motivation but do not possess the will to implement these ideas, perhaps, because of the risk of failure. Volition or will power does not eliminate the project risk but this is what determines a true leader of today's world.

Organizations with strategic intent to succeed must not merely look for motivated leaders but leaders who display volition to put innovation to the test.

4.4 THE CHALLENGE OF KEEPING A TEAM ENGAGED WHILE WORKING REMOTELY

It is true that almost all the leaders of companies at earlier times have worked with their teams partially virtually. However, no leader has had the experience of complete remote working, a situation caused by the pandemic. With the available data of the experience of remote work from home, it is clear that most of the employees are dealing with frustration. Leaders too are prone to frustration arising out of constant remote working. For a leader, the frustration is less because of working via virtual conferences and more because of the perceived productivity and expected performance of the team working from home.

While the introverted employees would be more productive with less of physical interaction with others, the extroverted employees become less productive in the absence of interaction with other human beings.

The most challenging role of a leader during the remote working situation is to strike the right balance between being empathetic while at the same time continuing to push on performance.

The challenge arises as the leader is in a dilemma about pushing team performance when every member of the team is either emotionally or mentally frustrated or distracted by the crisis.

Given this precarious situation, a leader must come to terms and accept the reality of remote working by placing trust in the team to perform. This is a great challenge to leaders who are extroverts themselves. Introverted leaders, since they perform better with less interaction with others, easily believe that their team too would perform to their best without frequent meetings in person or constant virtual monitoring. Given the lack of physical interaction among employees, the leader must balance the urge to "check on" productivity with "checking in" about their overall well-being.

> "Meet them where they are, both mentally and emotionally, rather than where you want them to be"

To ensure that the team is engaged, virtual conference calls are to be planned and designed to align the employees to the expected levels of productivity. This could take the form of daily virtual conference calls to bring everyone on to the same page of work plans. To enhance better social interaction even during virtual conference calls, the leader could delegate other team members to host the call.

Leaders must recognize that the team members need them to understand their anxieties and gently help them rebuild confidence before expecting high productivity and performance. The challenge here for the leader is to get to know each of the team members personally. Taking time out to listen to the team will put them in a comfortable space to engage personally with the leader. The leader must also consciously and generously appreciate the employees for different accomplishments.

4.5 CHALLENGES OF GLOBALIZATION TO TODAY'S LEADERSHIP

The life of modern-day leaders is more demanding now than ever because of globalization that has had enormous impact on today's leadership.

For leaders, globalization is going borderless in creating value for people. Globalization eliminates barriers to trade, communication and exchange of products, ideas, knowledge and culture. Leadership in a globalized context must excel in engaging all the resources required for success.

> For leaders, globalization is going borderless in creating value for people.

Globalization offers challenges to leadership that are both internal and external to the organization. Within the organization, the challenges for leaders are leading diverse groups of people, working across boundaries of the organization, improving performance and efficiency and propelling growth. This literally demands leaders with extra drive, unwavering energy, optimistic enthusiasm, enduring perseverance and openness to diversity.

External to the organization, the greatest challenge of globalization for leadership is leading beyond boundaries and barriers. In this age of globalization, leaders cannot limit their actions to the confines of an organization or territorial borders of a state or a country. A leader has to work across cultural boundaries, meet the requirements of different governments, navigate ahead of the global competitors, satisfy the expectations of the stakeholders and get work done from people who are very different one from another. As the business world is turning out to be more and more a global village, organizations and leaders need to think global, act global and go global.

4.5.1 Characteristic Traits of Leaders Challenged by Globalization

i) To rise above the challenges and reap maximum advantage of globalization, today's organizations need leaders who are able to understand and work in diverse cultures and geographies. The focus should be on building, fostering and establishing collaboration across boundaries.

ii) Today's leadership must possess the holistic perspective to understand that businesses and organizations need to change and evolve continuously as they are constantly affected by globalization and the ever changing social, economic and political environment. The utmost challenge then for today's leadership is to put the organization along "permanent transformation mode".

iii) Leaders leading teams in this time of globalization must possess the understanding of cultural appropriateness or awareness of how to interact and get things done in select cultures.

iv) To compete globally, organizations need leaders with the vision and volition to successfully benchmark their organizations against global organizations. This challenge of benchmarking against world class organizations implies managing sustainable growth, competing in global markets, effecting change management and developing future leaders.

4.6 THE CHALLENGE OF BURNOUT WITH TODAY'S LEADERSHIP

Burnout, a state of chronic stress, is one of the greatest challenges with today's leadership. It is a hazardous syndrome as it creeps into the life of high-achieving leaders even without their realizing it. Leaders who are passionate about their job usually take on enormous pressure and exceedingly heavy workloads. In pursuit of results, productivity and performance leaders who tend to work for hours on end without taking breaks gradually become victims of burnout. In a fast-paced culture that rewards "busyness", there is often pressure among leaders to move at breakneck speed without having time to slow down. Leaders live up to this challenge at a significant cost to their well-being resulting in burnout. While leaders are supposed to have strategies in preventing burnout among employees, it is rather common that leaders themselves are the most susceptible victims to burnout in organizations. Surveys reveal that, as a matter of fact, burnout among leaders is increasing at an alarming rate.

> "Leadership burnout is best characterized by emotional exhaustion that results in decreased productivity and performance in the workplace"

Burnout results in exhaustion, fatigue, detachment and feelings of ineffectiveness. It affects both the physical and the psychological health of individuals. Physiologically, burnout leads to chronic fatigue and insomnia. Psychologically, it results in the individual exhibiting pessimism and self-isolation. Leaders, when caught unawares with burnout, feel that they are no longer effective in their job and unable to face the challenges of leadership as they feel dejected, emotionally deprived with a depressing sense of meaninglessness. These eventually lead to lack of productivity and performance.

4.6.1 Common Causes of Burnout among Leaders

i) The most common cause of burnout is poor mental hygiene. Undue worrying, exaggerated overthinking and ruminating lead to tremendous energy leaks or drains. These unhealthy mental habits or traits decrease mental resilience causing people to feel overwhelmed and consequently burnout.

ii) One obvious cause of burnout is a work schedule without time for physiological rest or recovery. While organizations love leaders who "burn the candle at both ends", in the long term this is not a sustainable habit as it leads to mental fatigue and burnout.

iii) In an era bombarded with information overload, "digital overwhelm" gradually takes its toll on the cognitive abilities of leaders. The fast rate of input of information along with the increased volume of information come on as a bang vying for the attention of the

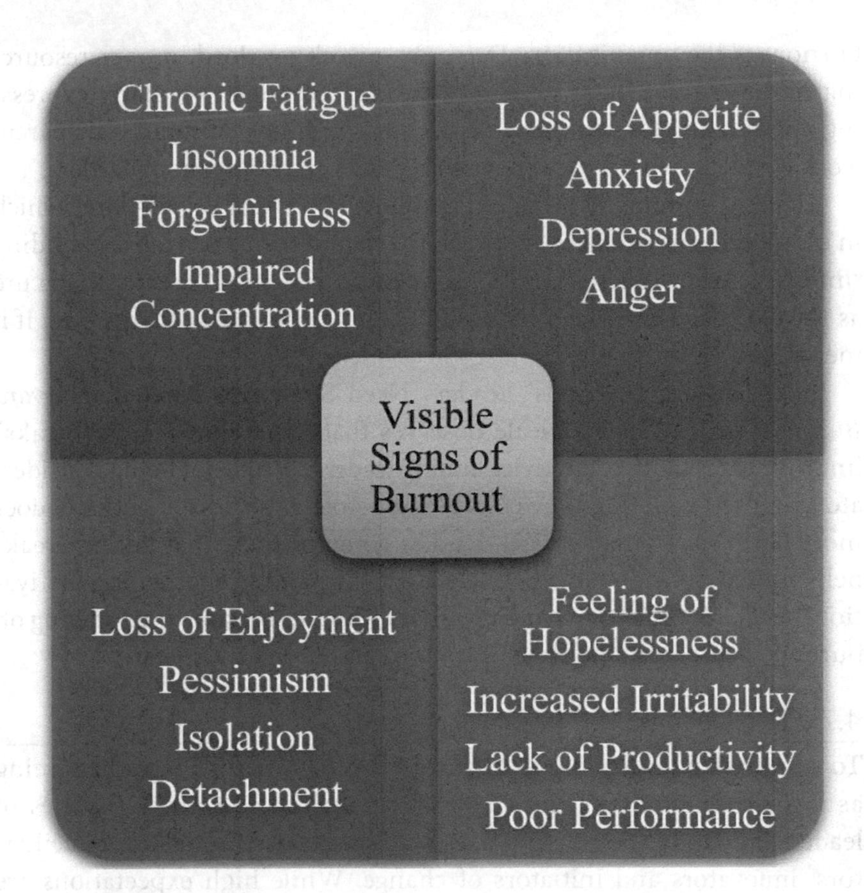

FIGURE 4.4 Visible signs of burnout in leaders.

leaders. This gives very little time for the leaders to rest and recover both physiologically and mentally leading to gradual burnout.

iv) The other common factors leading to burnout as indicated by various studies and journal articles include long work hours, increased responsibilities and duties, perceived malpractices in an organization, harassment, organizational politics and loss of work-life balance as the most common factors leading to burnout among leaders.

4.6.2 Preventing and Reducing Burnout

Reducing burnout among leaders is also a responsibility of organizations as much as it is a necessity on the part of leaders. Preventing the situations causing burnout should be priority goals as the ultimate casualty of burnout is the organization. Decreasing work overload, proper resourcing of duties, creating a value-based work ecosystem, regularly expressing appreciation for achievements and establishing a fair and transparent work environment are proven ways to reduce burnout among leaders.

The most commonly proposed antidote to burnout is "self-care" which in essence means balancing between the requirements of a job and finding time for oneself. When it comes to burnout, prevention is better than cure as the old adage goes. Leaders need to take time out to recharge even if it means just a few minutes amidst a hectic schedule.

In the article, "'Serious' Leaders Need Self-Care, Too", in *Harvard Business Review*, Palena Neale observes that even short "micro-breaks" improve focus and productivity among leaders. While not being considerate about self-care might be considered heroic or selfless, it actually does more harm than good. In fact, studies go on to show that taking breaks help prevent decision fatigue, lead to renewing and strengthening motivation, result in productivity and creativity while at the same time staving off burnout. Self-care actually helps leads to perform at their best.

4.7 CONCLUSION

To summarize, leadership has evolved through many phases beginning as a top-down function to a bottom-up reality. The many theories of leadership define the role of leaders as motivators, influencers, facilitators, innovators and initiators of change. While high expectations are placed on leaders by organizations, leaders are confronted with varied challenges. There are challenges that need to be addressed constantly like inspiring others, guiding change, developing new leaders. And then there are challenges that are time specific as was the case during the COVID-19 pandemic. While all organizations were affected including employees, leaders had to and still face the challenge to demand productivity while still keeping teams engaged working remotely. Some challenges have been caused by the worldwide phenomena of globalization where leaders are called upon to transcend beyond barriers and compete against world-class organizations. While some leaders are great motivators, they lack the willpower to implement innovative ideas because of the fear of failure. However, there are high-achieving leaders who go full throttle in benchmarking their organizations against world-class organizations, sometimes at the risk of their own mental health. The challenge for these leaders is not to fall victim to burnout that can drastically reduce production and performance.

BIBLIOGRAPHY

The American Review of Public Administration, Volume 36, Number 4, December 2006 382–391. © 2006 Sage Publications http://arp.sagepub.com hosted at http://online.sagepub.com

www.forbes.com/sites/niharchhaya/2020/03/23/here-are-the-top-five-leadership-challenges-during-the-coronavirus-pandemic/?sh=2879d67b5984

https://hbr.org/2020/09/preventing-burnout-is-about-empathetic-leadership?registration=success

https://hbr.org/2020/10/serious-leaders-need-self-care-too

http://integralleadershipreview.com/5401-feature-article-reframing-leadership-for-a-postmodern-age/

www.lead2xl.com/leadership-in-a-postmodern-age

www.psychologytoday.com/us/blog/high-octane-women/201311/the-tell-tale-signs-burnout-do-you-have-them

www.shrm.org/executive/resources/articles/pages/leaders-burn-out-too-moss.aspx

Sternberg, R. J. (2007). A systems model of leadership: WICS. *American Psychologist*, 62, 34–42.

CHAPTER 5

A Behavioural Approach to Leadership in Education

Preetha Mary George
Princy Sera Rajan
Bobby Mahanta
Zouhour El-Abiad

CONTENTS

5.1 Introduction 62
5.2 What Is Leadership Theory? 63
5.3 Traits of Good Leadership 63
5.4 Leadership Theories 63
  5.4.1 Great Man Theory 63
  5.4.2 Trait Theory 64
  5.4.3 Contingency Theory 64
  5.4.4 Situational Theory 64
  5.4.5 Behavioural Theory 65
5.5 Leadership as a Process in the Workplace 65
  5.5.1 Strengthening Your Strengths 65
  5.5.2 To Be an Inclusive Leader 66
5.6 Traits of Leadership in the Educational System 66
  5.6.1 Leadership Approach of Teachers in the Classroom 67

DOI: 10.1201/9781003244660-5

5.7 Attributes of a Great Educational Leader 67
5.7.1 Team Building 67
5.7.2 Positivity 67
5.7.3 Reflection 68
5.7.4 Execution and Planning of a Vision 68
5.7.5 Passion 68
5.8 Managers and Leaders 69
5.9 Conclusion 69
References 69

5.1 INTRODUCTION

Leadership skills are crucial to both teachers and leaders of any organization. Teachers are the leaders in the classroom. In this study we try to offer a reflection of educational leaders and leaders with exceptional leading skills. We try to answer questions on how leadership skills are important to educational leaders and what are the traits of a leader? Different leadership theories are dealt with in this work and our observations from real-life classroom situations with suggestions on how to build a leadership approach among teachers. Numerous articles on theories and its definitions already exist. This chapter provides a clear definition on different behavioural leadership theories.

Leaders with exemplary leadership skills are vital for all industries, businesses, and companies and for the educational institutions to be successful [1, 2]. The skill of a leader is pivotal for the positive outcome of a team. Without the presence of a leader no firm will run smoothly. They are part of effective practices. They won't rise up inherently. That's why many management institutions run degree programs to train and help students to become great leaders.

In the world of leadership there exist many theories on leadership, its working style, how to make good leaders, and how to be an effective leader. The leadership theories explain how leadership styles work for a business, or an educational institution to bring success [3]. These theories explain how leaders possess and develop leadership traits. If you wish to become a successful business leader, it's important to know the leadership theories, leadership styles and how it affects the management and leadership style. According to Ralph Nader, "the purpose of leadership is to bring out the leaders and not the followers". A leader is someone who can change the perspective of a follower and change them to be a leader [4]. An inspiring leader always elevates and inspires the public.

5.2 WHAT IS LEADERSHIP THEORY?

What makes leaders? Leadership theory explains how to become leaders, what makes certain people become leaders. Leadership theories encourage followers to become leaders. The study of leadership theory helps to find out why some people are selected as leaders and manage while some remain followers.

5.3 TRAITS OF GOOD LEADERSHIP

Traits are vital to leaders. A good leader should possess:

- High morals and strong ethics
- Self-organizational skills
- Steadfast leadership
- Nurture the growth of employees
- Foster social connections

5.4 LEADERSHIP THEORIES

5.4.1 Great Man Theory

The *great man theory* came into existence at a time when leadership pertained to males. In the early days, leadership was considered to be a quality inherited by males, so it was named Great man theory. But later by the emergence of great women leaders, this theory was reconciled as great person theory. Great man theory states that leaders are born with the right characteristics and qualities like charm, intellect and communication for leading a team [5]. This theory further says that the world's great leaders are born and not made. This theory focuses on the innate qualities of an individual. It says a leader can rise to any situation when the need soars. It implies that the people in power deserve to lead because of endowed traits. It exposes that leaders are born not made. It describes leaders as fearless, whimsical and determined to rise to a leader when a situation arises.

5.4.2 Trait Theory

Trait theory is close to the great man theory. This theory is used to anticipate effective leadership. The identified qualities with physiological attributes like appearance, body mass index and statistics such as age, education, family background, intelligence are compared to prospective leaders to determine their leadership success. This theory focuses on the qualities of intelligence, integrity, sociability and determination. Trait theory focuses on innate qualities that makes a person a leader. This theory advocates that leaders and their traits are the key to the success of an organization and focus on the leader and avoids the follower.

5.4.3 Contingency Theory

Renowned researchers Hodgson and White say that effective leadership is one that can find an ideal balance between behaviour and need. Good leaders inherit the perfect qualities to analyse the needs of their companions. In short, *contingency theory* advocates that a significant leadership is an amalgamation of many principal variables. This theory can be categorised into (a) environmental contingency and (b) internal contingency. Environmental contingency depends on the stability of the environment. Internal contingency focuses on the size of an organization. Contingency theory holds good on situational variables such as leadership style, job design, involvement in decision making etc. It varies in opinion of the orientation of the individual. Preference of a leader plays a significant part in the ability to be successful in different situations. This theory balances between the preference of an individual and the situational factors.

5.4.4 Situational Theory

This is very close to contingency theory. As the name proposes, *situational theory* hints that leadership hinges on the situation at hand. According to Paul Hersey and Ken Blanchard, this theory combines two key variables, the style of the leadership and the follower's level of maturity [6]. It suggests that no single leadership style is best. According to this theory, effective leaders are those who adapt their leadership style to the situation and look at the factors to get the task completed. Instead, it depends upon the suited strategies. They classified *maturity* into four stages.

M1: The team members do not hold motivation or the prudent skills to achieve tasks on time.

M2: The team members are willing and ambitious to do work, but do not possess the necessary ability.

M3: Members of the team are not willing to be responsible for accountability, but they hold skills and capacity to achieve tasks.

M4: Members of the team are equipped with all skills and possess talents and are motivated for the project's achievement.

According to this theory, a leader executes leadership established on the above levels of maturity of their team.

5.4.5 Behavioural Theory

Behavioural theory focusses on the characteristics and efforts of leaders rather than their attributes. It is implicit that successful leadership is the outcome of the skills learnt. A leader needs three attributes to lead their followers: scientific, conceptual and interpersonal skills. Technical skill refers to the knowledge of a leader in adapting to any process or technique [7]. Human skills help leaders to relate with other humans. Conceptual skills help the leader to run the institute/organization steadily.

5.5 LEADERSHIP AS A PROCESS IN THE WORKPLACE

Leadership theories help to improve an individual's leadership skills. It is a process of motivating others to achieve a goal. It is a set of behaviours that can be learned from a set of skills that can be nurtured. It is bound to emotional commitment. It affects how others think about work. It is an important fact in management. It is a process that can be observed, trained and learned. It cannot be defined as a role. It is a relationship between leaders and followers to achieve goals.

5.5.1 Strengthening Your Strengths

Successful leadership is based upon the qualities of a leader. Leaders concentrate on their strengths rather than their weaknesses. The positive energy differs from person to person. A strong will is key for a leader. A strong-willed leader finds their own inner strengths and overcomes the challenges they face to tackle a situation. A decisive nature is another quality a leader should possess. When the followers are perplexed to make a decision, the leader should peacefully evaluate the circumstance and choose the correct steps to unify and lead everybody. Even if the situation goes wrong, they must be willing to rise and awake from their mistakes.

5.5.2 To Be an Inclusive Leader

An inclusive leader always allows fellow colleagues to be trained as leaders by receiving the feedback of the others or dispensing more responsibilities to others. An inclusive leader is one who is aware of one's own biases and preferences. An inclusive leader adapts quickly to diverse situations. Inclusive leaders make an attempt to recognise people for their own work. Inclusive leaders will motivate and elevate others. An inclusive leader must be an active listener and seek different reflections from their team members. Inclusive leaders will communicate with their team with an open mind.

5.6 TRAITS OF LEADERSHIP IN THE EDUCATIONAL SYSTEM

An educational system also requires strong leaders with exemplary skills. A teacher is an invaluable leader who tries to create a better community environment. Teachers are the heart of the classroom [8]. Educational leaders drive students, teachers and even schools to success. Educational leaders should grow in practical skills and always evaluate and improve their own leadership competencies. An effective educational leader embodies the following traits:

- *Visionary*: A visionary is a person who has a better outlook for the future.
- *Passion*: Passion is a strong desire to do a particular activity.
- *Respect*: Giving respect and taking respect are good moral gestures which can be inculcated by all.
- *Honesty*: The moral character of a human being reflects the truthfulness of a person.
- *Integrity*: A character with high moral principles.
- *Courageous*: A person who has a strong mind to face difficult and different situations.
- *Dedication*: A person who is dedicated and has a passion for an idea.
- *Compassion*: A feeling to help others.
- *Communicator*: A communicator is a person who communicates to others in a good and effective manner.
- *Influence*: A person who affects other people's thinking.

5.6.1 Leadership Approach of Teachers in the Classroom

Leadership skills of teachers to build enjoyable learning are:

- A good teacher is a good listener
- A good teacher has empathy for their students
- A good teacher eliminates the stiff boundary between a teacher and a student
- A good teacher builds self-confidence among students
- A good teacher builds close relationships between a teacher and a student
- A good teacher is a shadow to shed thc feelings and thoughts of students

5.7 ATTRIBUTES OF A GREAT EDUCATIONAL LEADER

Leadership is nothing more and nothing less. Leadership can influence a crowd. Being a leader is not that easy. Leaders must possess effective qualities to achieve goals and success. To Mahatma Gandhi, it was resistance and persistence. To Elon Musk, it was the vision. To Warren Bennis, it was the capacity to transform a vision in to reality. Successful great leaders have unique qualities.

5.7.1 Team Building

Team building is a critical quality of an educational leader. By effective teaching a leader can reach students, parents and management to unite for the success of a goal. Team building is a process of helping a group to become more efficient in achieving a task. It can be considered an organizational development strategy. It is a strategy to create a greater impact to expose an organizational team. Teamwork ability is an ability that every worker must acquire for any job in any organization. Employers and managers always look for the individual who can work as a team.

5.7.2 Positivity

A strong educational leader can overcome the hurdles faced by the school to solve an issue. Moreover, these leaders communicate to the management of the school or institution and are inspired by everyone to overcome all the obstacles. This quality is the grit or perseverance of a leader. Positivity enhances work performance of employers, managers as well as employees. It is a state of mind that envisages any situation favourable to execute a plan. A person can change their life by thinking positively about things. If a person is positive, their actions portray positive emotions and they engage in positive characteristics like generosity and kindness. All this positivity helps to ensure positive outcomes and ultimately helps in having good mental health and well-being.

5.7.3 Reflection

Effective leaders reflect on themselves and their growth. This reflection is a drive to look back and reveals what plans need to be taken for the further operation. It is analysing one's own experience to improve the way they work, to find flaws etc. It is related to the theme of learning by yourself. In this process, you will think about what you did, what went wrong and how you can improve such situations in future. Once a leader adapts to the reflective process, they will find it very useful at work and to manage their team. It is a bridge to that gap between theory and practice. It is an ethical set of skills in real time dealing with complex situations. It encourages us to explore ourselves to find solutions to problems.

5.7.4 Execution and Planning of a Vision

A leader should be clear in their vision, actions and thoughts. Execution planning can also be called project planning or prior planning. It is creating a strategy to accomplish a task. It can be considered a first task in project management. It is detailed planning to identify the strength and resources. Planning and execution are interconnected. Before the start of any project, an action plan must be envisaged to implement that work. The hardest part in strategy is not planning but how to execute it. During this stage, project managers and leaders sketch and design an idea with an execution of plan.

5.7.5 Passion

Leaders should be driven by passion [9]. Passion is an inherent interest in something. It leverages doing things amazingly. It encompasses skill and talent and how they work together. It is subjected to change and evolve. It may be long- or short-lived. The key to achieving anything is to try without stopping. A passionate leader cares about their followers. Communication, passion, authenticity, team building reflection is the assay of an educational leader to change the lives of teachers, students, parents and management.

5.8 MANAGERS AND LEADERS

Managers and leaders are different. Managers search for stability and control, they try to solve problems instantaneously [10–11]. The manager manages the organization while the leaders lead a team and sway team members to achieve a goal. The manager shows organizational, management and problem-solving skills. Managers lead by authority. Managers require many people to work efficiently. Leaders tolerate chaos and won't want to solve issues instantly. A leader inspires and influences. Leaders always take care of their team. Leaders have a strategic view. Leaders are open minded and promote innovation. They are willing to delay the closure of issues. But organizations and institutions need both leaders and managers. Leadership is practical to direct affairs.

5.9 CONCLUSION

Great men were born, not made. A leader is one who has innate leadership qualities. Leadership can be democratic, autocratic or laissez-faire. The autocratic leader makes their own decisions. Democratic leaders consider the decision of all. Laissez-faire assumes the role of a leader. Leadership is a series of personal traits. From the study of leadership styles, a leader is one who motivates, guides and monitors their employees. From this study, a leader is one who can mobilise followers. A leader is enthusiastic and possesses good communication skills. They will be dominant over a group and teammates and should influence employees in an organization.

REFERENCES

[1] K. Kadri, A. Mansor, and M. Nor, "Principal and teacher leadership competencies and 21st century teacher learning and facilitating practices: Instrument development and demographic analysis," *Creat Educ.*, vol. 12, pp. 2196–2215, Sept. 2021, DOI: 10.4236/ce.2021.129168

[2] Van Dierendonck, "Servant leadership: A review and synthesis," *J Manag.*, vol. 37, pp. 1228–1261, July 2011, DOI: 10.1177/0149206310380462

[3] Asep Suryana, Widiawati Widiawati, and Minnah El Widdah, "Leadership approach: Developing teacher leadership skills in the classroom," *Proc 3rd Int Conf Res Educ Adm Manag.*, vol. 400, pp. 381–386, Jan. 2020, DOI: 10.2991/assehr.k.200130.205

[4] Immanuel Yosua, Juliana Murniati, and Hana Panggabean, "The meaning of leadership for leaders of private universities in Indonesia," *Jurnal Ilmu Administrasi dan Organisasi*, vol. 28, pp. 133–142, Sept. 2021, DOI: 10.20476/jbb.v28i3.1284

[5] J. York-Barr, and K. Duke, "What do we know about teacher leadership? Findings from two decades of scholarship," *Rev Educ Res.*, vol. 74, no. 3, pp. 255–316, Sept. 2004, DOI: 10.3102/00346543074003255
[6] M. W. Berkowitz, and M. A. Hoppe, "Character education and gifted children," *High Abil Stud.*, vol. 20, pp. 131–142, Dec. 2009, https://doi.org/10.1080/13598130903358493
[7] A. Fallis, A. Madkur, D. Narvaez, D. Lapsley, A. Silvia, M. M. Steedly, and K. Larson, "Understanding the importance of character education," *ACLL.*, vol. 3, no. 4, pp. 1–16, 2013, https://doi.org/10.1017/CBO9781107415324.004
[8] Dohyoung Koh, Kyootai Lee, and Kailash Joshi, "Transformational leadership and creativity: A meta-analytic review and identification of an integrated model," *J Organ Behav.*, vol. 40, pp. 625–650, July 2019, https://doi.org/10.1002/job.2355
[9] H. Thamhain, "Changing dynamics of team leadership in global project environments," *AJIBM*, vol. 3, no. 2, pp. 146–156, DOI: 10.4236/ajibm.2013.32020
[10] R. B. Santos, E. O. de Sousa, F. V. da Silva, S. L. da Cruz, and A. M. F. Fileti, "Detection and on-line prediction of leak magnitude in a gas pipeline using an acoustic method and neural network data processing," *Braz. J. Chem. Eng.*, vol. 31, no. 1, pp. 145–153, Mar. 2014, DOI: 10.1590/S0104-66322014000100014
[11] R. P. da Cruz, F. V. da Silva, and A. M. F. Fileti, "Machine learning and acoustic method applied to leak detection and location in low-pressure gas pipelines," *Clean Techn Environ Policy*, vol. 22, no. 3, pp. 627–638, Apr. 2020, DOI: 10.1007/s10098-019-01805-x

CHAPTER 6

Key Elements That Bind Leadership with AI

Bharath Reddy
Palak Goel
Vasireddy Bindu Hasitha
Anil Audumbar Pise

CONTENTS

6.1 Introduction 72
6.2 Aids in the Prediction 72
6.3 New Insight Requires New Initiatives 72
6.4 Computer-Based Intelligence Is Making Wins No Matter How You Look at It 73
6.5 Using AI's Predictive Power Will Necessitate Best Judgement 73
6.6 AI Is Susceptible to a Few Human Faults 74
6.7 Humans Are Unlikely to Be Replaced by Machines Very Soon 74
6.8 What Can AI NOT Mimic? 77
6.9 Conclusion 78
References 78

DOI: 10.1201/9781003244660-6

6.1 INTRODUCTION

Leadership is defined as the capacity to delegate, inspire, and motivate others in one's organization, as well as the ability to handle challenges in a constantly changing environment. Artificial Intelligence is predicted to transform the workforce dramatically. The discussion is currently at a fever pitch, ranging from automating ordinary operations to redefining client experiences. We look at how artificial intelligence (AI) may assist senior leadership in making strategic decisions, which are at the heart of every successful business. Senior managers and developing business leaders in firms, whose time is consumed by extremely operational tasks, require AI to free up time for human-certified jobs. Making decisions is crucial. Even seasoned leaders with a demonstrated history of working on solid judgement might be perplexed by a difficult scenario. Leaders are keen to understand if Artificial Intelligence (AI) can genuinely make a difference for them, since Artificial Intelligence promises to create a profound revolution across enterprises and sectors [1].

6.2 AIDS IN THE PREDICTION

By analysing historical records, predictive analytics will help a corporation predict future occurrences and patterns. Furthermore, machine learning is at the heart of AI, since every platform, programme, or application acquires insights from incoming data, adjusts to developing patterns, and progressively begins to respond to emerging trends. This benefits leaders in various ways:

- The capacity to evaluate cost-effectiveness and prospective return on investment.
- Through continuous customer behaviour research, the possibility to optimise the buyer's journey exists.
- There is enough leeway to accurately assess the success percentage of established targets.

6.3 NEW INSIGHT REQUIRES NEW INITIATIVES

Any discussion encompassing AI requires a re-appraisal of administration. The responses lie somewhere else, in a change from shrewd, to insightful authority. For shrewd pioneers don't just make and catch fundamental financial worth, they likewise assemble more feasible—and genuine—associations. More than sensible, they are mindful, respecting their trustee obligation of dedication, and care to the association and its feasible/long-haul esteem. Esteem that AI can improve, disintegrate, or obliterate, contingent upon how astutely it is driven.

Basically, astute independent direction is tied in with widening the context-oriented structure and giving a more comprehensive point of view. It implies having the option to comprehend and resolve inconsistencies, conundrums, pressures. Savvy pioneers embrace a 'multifield viewpoint' and have the enthusiastic development and liberality of soul to motivate and assemble others. Man-made intelligence can assist pioneers with emerging an association's vision, however without astuteness, it might jeopardize a more sympathetic future [1].

6.4 COMPUTER-BASED INTELLIGENCE IS MAKING WINS NO MATTER HOW YOU LOOK AT IT

Computers process enormous amounts of data to detect patterns and make predictions in Artificial Intelligence. Deep Learning permits prescient displaying by means of counterfeit 'neural networks'—freely demonstrating the way neurons associate in the mind.

6.5 USING AI'S PREDICTIVE POWER WILL NECESSITATE BEST JUDGEMENT

Any leader understands the value of foreseeing the next market change or excellent product while juggling several variables. In terms of rising above the noise and capturing patterns and signals, Artificial Intelligence is already exceeding humans. Its accurate forecasting power complements and improves human judgement. However, utilising Artificial Intelligence necessitates altering essential operational tasks and working across internal organizational boundaries to create an ecosystem of shared virtual connections and practises. Data analytics must be entrenched as a key organization strategy, used to discover pain spots, design solutions, and empower decisions, in order for predictions to lead to strategically sound actions.

Business is all about imagining the impossible, developing innovative solutions, and motivating and mobilising others. Rather than just anticipating specific occurrences, shaping a future environment. Wise leaders will blend human and artificial intelligence, using AI to its full potential. Shrewd pioneers use their inventiveness where they can have an effect, by building up another arrangement they can impact. Man-made intelligence is coming to a long way past Industry 4.0. It has been changing monetary administrations for quite a long time, illuminating venture choices.

6.6 AI IS SUSCEPTIBLE TO A FEW HUMAN FAULTS

Assuming that AI is intended to copy the human mind, then, at that point, it is likewise error prone—liable to inclination. It will in general limit the chance of critical change, working inside the world characterized by the information used to adjust it in any case. Artificial intelligence, and especially Machine Learning, should be cleaned up, observed, and oversaw by savvy, mindful pioneers, with information respectability protected, the right information inputs, and certain, versatile calculations.

6.7 HUMANS ARE UNLIKELY TO BE REPLACED BY MACHINES VERY SOON

For a long time to come, AI may drastically adjust how work finishes, supplementing and increasing human capacities. Intellectual frameworks can perform explicit undertakings, turning out to be cannier constantly through input circles. Astute initiative will recognize the gigantic open doors and ability of PC learning, illuminated by bits of knowledge from neuroscience, while stressing human imagination. Not endeavouring to contend with PCs, however, fostering our human characteristics—innovativeness, wisdom, decency of judgment, social coordinated effort, and an all-encompassing vision of things to come.

Through Artificial Intelligence, the following can be achieved:

- Embrace and develop the joint effort among human and AI: Changing activities, markets, enterprises—and the labour force—with new abilities.
- Imagine a more significant future: Show authoritative partners what it can resemble, and guide and empower their association to seek that objective.
- Subsequently, ingrain productive headway while making society a superior spot to live maintaining their guardian obligation to the association, its portion and partners, and the local area overall.

The future 'intellectual organization' will appear to be exceptionally unique to anything we know today. However, information is only a lot of numbers that are inane without setting. So, AI-wise pioneers ought to work with advancement, accepting joint effort among human and Artificial Intelligence, changing tasks, markets, enterprises—and the labour force—with new abilities.

It is well known that AI is being used to help experts across all industries to diagnose and solve problems much faster, thus also enabling consumers to do confounding tasks and build a more comfortable environment for themselves [2]. With so much happening, it is tempting to regard artificial intelligence as a 'threat' to human leadership. The purpose of AI is to augment and improve. Instead of eliminating jobs, AI can help existing employees to evolve into new positions. The need for human interaction will outweigh the need for cutting head counts. Today's more innovative leaders are preparing now to work alongside AI, striving to seek answers to questions that weren't previously economical to answer, as well as impose questions they didn't even know how to frame. To understand how AI can change a workforce, it is essential we begin with discussing the key elements that bind AI with Leadership [3].

- The past decision makers of an organization spent a considerable amount of time on administrative tasks such as scheduling, coordination, and step-by-step decision making for small tasks. AI will be able to do this and other complex jobs with better competence and productivity. Managers and people in leadership positions need to now understand the technical role of Artificial Intelligence and acquire force to fill in the gaps where AI is far from being developed, such as interpersonal and soft skills. Decision making binds AI with leadership and can help an organization reach better conclusions. When leaders are able to eliminate busy work, decision fatigue, and become more visible and accessible, the entire company benefits. While some aspects of decision making still include human intelligence such as bridging the gap between technology and people, employee development, maybe even the recruitment process in a company, etc. Henceforth, it is crucial for managers and leaders who will vanguard the transition of the current workforce into automation regulated archetypes, to possess the right set of skills to balance and manage both the machine and human side of decision making [4].

- Change will most definitely begin with the leader, in any organization or a company, or a task force. It is in the hands of the leader to embrace AI and set forth an example for others in their organization to utilize the many marvellous advantages of AI. By automating tasks, they can free up their time to work on strategic initiatives, work with people on one-to-one basis and thus help them optimize their skills. This not only opens doors for greater efficiency and growth but allows deeper connections and insights, helping the leaders realize the full potential of their workforce.

- With the right data, AI can be used to monitor performances, and help an organization to plan and span the execution of a bigger picture within its reach. It can be used to gain a better understanding of how work is done in a company, map out which employees have the greatest impact and where there might be talent gaps. Such intelligence ultimately brings together the determination, discipline, perseverance, awareness, and accountability into any task force. It strengthens quality and increases the chances of creating a more promising outcome.

- As mentioned before, there exist algorithms like Machine Learning and Deep Learning which allow software applications in predicting outcomes without being explicitly programmed. Machine learning models like supervised and unsupervised learning etc., are pivotal in helping intelligent machines realize the improvement of the performance of a system beyond what it can provide by other analytic techniques. For example, with prediction techniques, AI can help leaders enter new markets more seamlessly, as it has the niche to inform leaders whether a geographical area's talent and business conditions offer a viable opportunity for expansion or not. So, to build on this point, it is advantageous for an organization (such as businesses and the public linked with say, stock markets) to know what to expect based on a promising prediction system.
- AI can help leaders to synthesize their organization's vision, but without proper guidance, a more humane future is at the stake of being endangered. It is essential for the leaders to know, that adaptability to change is the key. So, leaders of the new age must embrace the collaboration between a machine and a human. While adaptable algorithms are required in automation to not reproduce any system bias, a type of adaptability in humans is also required to fill the gaps that a machine can fail to fill. At an organizational level, adaptability

means being ready to welcome innovation and to innovate, respond to opportunities and threats as they appear. Not to forget, a team of individuals equals to an organization. “Adaptability” at an individual level means being open to new ideas and being allowed to change and influence change in opinions relevant to a subject. In the AI age, it is necessary to consider every perspective and create a balance for the benefit of a society [5].

- Leaders must treat artificial intelligence the way they would treat new employees joining the company: trust them, little by little; watch the outputs and coach/adjust as they move forward. It is crucial for every leader using AI to understand that “Artificial Intelligence” is not a magic elixir. It cannot fix all problems. It cannot be blindly trusted as it will create algorithmic bias or lead to systems without the oversight that's necessary to ensure that the algorithms perform optimally [6].
- With trust and time, AI will have a positive impact. To be more successful than others, leaders running organizations shouldn't lose the confidence to take risks and be willing to test new processes and exploring new technologies. Gone are the days when leaders could solely rely on ‘gut instincts’ to make key business decisions. AI can be used as a tool, not as a suitable replacement of an entire leadership and an organization that will follow this limitation, shall thrive.

6.8 WHAT CAN AI NOT MIMIC?

- It goes without saying that no matter the technical skill and predictive niche AI can provide, there are some qualities it simply cannot mimic. Human leaders have a significant edge over artificial intelligence in a few areas that make them indispensable to an organization. Human leaders can leverage the following:
  - *Creativity*: AI can induce logic into problem solving but is it the best way, all the time? Creative leaders who use their creative skillset, ideas and intuition to solve problems find better paths to move forward, because they have the compassion to understand how decisions and their consequences affect others in the company and its future.

- *Human touch*: AI is perfectly answerable for questions to which the answer is either yes or no (or a set facts regardless of context). But problems may still need a human connection to resolve.
- *Mission and vision*: As work becomes more automated, human values become more important. Mission and vision are two factors that inspire and motivate employees, and they cannot be replicated by machines [7].

6.9 CONCLUSION

This chapter discusses leadership before artificial intelligence and after artificial intelligence. It firstly explores the importance of leadership and functions of leader in one's organization. Then it focused on the importance of AI and then after how mission learning made lot of changes in AI. Then, it tells how efficiently leadership is working after the AI leadership comes and how the AIML made changes in leadership. Continued with how the feature improvement of AIML brings the changes in leadership. It also focused on how the level of feature prediction of market is changed after AI comes compared to human. Finally, this chapter tells about the key elements of AI leadership and how to collaborate human with mission like decision-making capability and work efficiency, it strengthens quality and increases the chances of creating a more promising outcome. At last, it relates that anything is possible by working with AI, so take a risk with AI for future growth of an organization. The methodology proposed in this chapter will work with the advancement of man-made brainpower, authority science, the board science, as well as big data analytics.

REFERENCES

[1] Zhaohao Sun, "Artificial Leadership: An Artificial Intelligence Approach," 2018. www.researchgate.net/publication/327105932_Artificial_Leadership_An_Artificial_Intelligence_Approach (accessed Jan. 14, 2022).

[2] Bob Friday, "4 Key AI Concepts You Need to Understand," *Infoworld*, 2017. www.infoworld.com/article/3200790/4-key-ai-concepts-you-need-to-understand.html (accessed Jan. 14, 2022).

[3] Great Learning Team, "Top 10 Leadership Skills You Need in the Age of AI," www.mygreatlearning.com/blog/effective-leadership-skills-you-need-in-the-age-of-ai/ (accessed Jan. 14, 2022).

[4] P. Verhezen, "Wise Leadership and AI Why New Intelligence Will Need New Leadership," 2019. https://www.amrop.com/news-insights/articles/wise-leadership-and-ai-new-intelligence-new-leadership/

[5] P. D. Steven T. Hunt, "2 Ways Artificial Intelligence Is Redefining Leadership," *SAP*. www.sap.com/india/insights/hr/2-ways-artificial-intelligence-is-redefining-leadership.html (accessed Jan. 14, 2022).

[6] "AI Can Never Replace Leadership. Here's Why," *Censia*, 2021. www.censia.com/blog/ai-can-never-replace-leadership-heres-why/ (accessed Jan. 14, 2022).

[7] M. W. Tomas Chamorro-Premuzic, "As AI Makes More Decisions, the Nature of Leadership Will Change," *Harvard Business Review*, 2018. https://hbr.org/2018/01/as-ai-makes-more-decisions-the-nature-of-leadership-will-change (accessed Jan. 14, 2022).

C. F. Gray, and E. W. Larson, *Project Management: The Managerial Process* (5th ed.). Boston: McGraw-Hill, 2011.

C. Howson, R. L. Sallam and J. L. Richa, "Magic Quadrant for Analytics and Business Intelligence Platforms," 26 Feb 2018. (Online). www.gartner.com (accessed Aug. 16, 2018).

B. L. Simmons, "Ten Most Important Leadership Functions," Sept. 19, 2010. www.bretlsimmons.com/2010-09/ten-most-important-leadership-functions/ (accessed June 19, 2018).

CHAPTER 7

How Does AI Leadership Affect Strategic Implementation

Amogh S. Jajee
Anushka Johari
Debdutta Choudhury
Daya Shankar
Dheeraj Anchuri
Jorge A. Wise

CONTENTS

7.1 What Is Strategy? 82
7.2 What Is Strategic Decision Making? 83
7.3 Introduction to Artificial Intelligence (AI) 84
7.4 What Is AI Leadership? 85
7.5 Impact of Quantum Computing on AI 89
Bibliography 92

DOI: 10.1201/9781003244660-7

7.1 WHAT IS STRATEGY?

Any organization, irrespective of its products or services or markets need a strategy. Approaching a problem without having a strategy is like shooting an arrow without aiming. Strategy is what gives purpose and focus to an organization.

What exactly is strategy?

The term *strategy* has a Greek origin and was usually referred to as the planning and directing of military missions during the war.

In today's world, the word strategy has been overly used in the corporate setting and is usually seen as something very complex. But strategy is a very simple and straightforward term. It doesn't have a particular definition but for general understanding, we can say a strategy is "A well-planned course of action to move from where we currently are to where we want to be in future".

A strategy is not just something that we do once and keep following it for all the problems or even similar problems in different situations, strategy is more of a continuous process and it has to evolve as the variables in the environment change.

Let's have a look at some real strategies from the past, during World War II, Hitler was on a complete rampage and it was important for the United States to break that streak. A group of 1,100 men officially known as 23rd Headquarters special troops were deployed, but the surprising aspect is that they were not soldiers, rather group consisted of actors, sound technicians, painters, press agents, photographers, and makeup artists; inflatable tanks, radio transmissions and sound trucks to stage 20 battlefield deceptions were used to create a psychological impact which was crucial for the allied victory that followed.

One of the recent crazy strategies in the automobile industry has been Tesla. It was well known that Tesla wanted to be a mass manufacturer of electric cars and when a new company wants to enter the market it tries to enter with an MVP (minimum viable product) where a company introduces a product with basic feature sets at a low cost and tries to penetrate the market, but Tesla did quite the opposite by introducing Roadster which was a fully loaded electric supercar which retailed over US$2,00,000 for the basic model. By doing this, the company gained a lot of attention and proved that electric cars could match or in some cases even surpass the power of traditional supercars, the company itself doesn't seem to be doing a lot of marketing but Elon by his personality and ambitious goals grab a lot of attention to his companies and today we can see that Tesla is undoubtedly the best electric car manufacturer in the world.

Even in a recent launch of the cyber truck event while showing off how tough the truck is, they threw a heavy steel ball at the window, which obviously broke the glass and Elon casually said, "we have room for improvement" and continued the event since it attracted huge publicity.

It is quite possible that this was completely planned since being in the news also ensures free publicity and creates awareness about the product which is a part of marketing strategy.

7.2 WHAT IS STRATEGIC DECISION MAKING?

Before understanding strategic decisions let's define what a decision is. In simple terms, making a decision is choosing an alternative or an act of choosing a particular course of action. Strategic decision making is also similar but with a tiny bit of mutation which makes a huge impact on the output; making decisions strategically essentially means that the decisions are made with keeping the long-term goal in mind.

Every business has a big goal to achieve and it is not possible to achieve it in one go. First, the goal has to be divided into smaller achievable bits, and then step by step we move towards the final goal while taking a strategic decision, the mission and vision of the organization are kept in mind and the decision for short-term steps is made in such a way that the company moves towards its end goal.

Let's understand this with an example, assuming a baby food manufacturer wants to create a brand image that he is the top-quality manufacturer, becoming top in terms of quality is the goal and every business decision he takes henceforth must be based on achieving that goal. For manufacturing the product various ingredients are required, the price of the ingredients will differ based on the quality, here the business will choose the higher-quality ingredients even though it makes the manufacturing process expensive because the final goal is to become a manufacturer that offers the best quality baby food; making this decision will determine entire future of the company. Even other decisions like which retail outlet the product will be available, the price of the product, the target market, and all other decisions will have to be made with the mission and vision in mind.

If we observe the strategic implementation of the decision, it is recognizing the mission and vision of the organization. It might be anything like capturing a particular percent of market cap, or achieving a certain amount of profit or working for a cause, and then taking that mission and vision and creating quantified long-term goals from them and based on those long-term goals creating short-term goals to break the process down into achievable steps and finally making decisions based on keeping that mission and vision in the picture.

We have understood strategic implementation in its simplest form and for small-scale businesses it is simple, we need to understand that all the decisions that are made are based on certain data, but as the business grows so does its target market and its reach and hence the amount of data that has to be dealt with also significantly increases and after a certain point it becomes extremely challenging for anyone to handle the data and create insights from of it, without the use of technology. This is where AI comes into the picture, AI can handle a huge amount of data and can create useful insights for the business which intern will help in the strategic decision-making process.

7.3 INTRODUCTION TO ARTIFICIAL INTELLIGENCE (AI)

Can intelligence, which we often associate with human beings, be taught to a machine? A mere computer or a robot controlled by a computer displaying intelligent traits is not a new sight. Artificial Intelligence, as it is called is the ability of a computer to perform tasks that human beings do. AI is essentially a branch of computer science that emphasizes and deals with machines learning to think and act by themselves.

The last few decades have seen tremendous progress in Artificial Intelligence. The term AI coined in 1956, has been defined by several experts according to their perception. John McCarthy in 2004 defined AI as "the science and engineering of making intelligent machines, especially intelligent computer programs. It is related to the similar task of using computers to understand human intelligence, but AI does not have to confine itself to methods that are biologically observable".

AI has made progress from the time when Alan Turing, the father of computer science asked the immortal question "Can machines think?" AI applications have transcended all walks of life starting from drug discovery, space applications, financial predictions, marketing decisions to even predicting which music to play depending on a person's mood and the time of the day.

AI owes its origins to the humble concept of data-based decision making. Data Analytics has taken off over the last decade and a half due to the development of several statistics-based analytical tools and techniques and automatic data availability due to automation and digitisation of businesses. Analytics along with machine learning algorithms are able to interpret data and predict outcomes. This concept forms the basis of what is known as artificial intelligence.

AI has applications in all aspects of business value chain be it customer experience or internal financial and operational decision making. The advantage of AI is an ability to predict based on past data and it gives the data oriented objective input any strategic decision maker would cherish. Coupled with human instinct, this data-based predictive input helps to make much more robust decisions.

7.4 WHAT IS AI LEADERSHIP?

In businesses and corporations, the concept of Human Resources is indispensable. R. C. Davis defines an organization as, "a group of people who are cooperating under the direction of leadership for the accomplishment of a common end". Thus, AI Leadership essentially means that for any organization, where the workers, directors, management, or anyone who is in a position where their decisions affect the course of business, they use Artificial Intelligence to aid their decision making for efficient results. The decisions made by humans are a result of numerous factors which range from the way they were brought up, their socialization, education, skills, experiences to even their emotions and mood; whereas when we leave the process of decision making to AI, the outcome is purely based on facts and how the machine's algorithm works. At the executive leadership level, especially, AI is making a lot of changes with its implementation where humans and machines can work together harmoniously. Together, because of the pace at which current businesses are moving towards complete digitalization, we need AI to augment our decision-making process.

According to a recent Infosys study titled "Leadership in the Age of AI", 45% of organizations globally experience AI deployments currently outpacing the activities of humans in that organization by a far greater margin. Therefore, it is with great care that AI should be integrated into organizations; which does not mean that companies should be reluctant to deploy AI for their activities, rather AI should be used to complement human progress and not compete with it. AI can be used in many different areas in businesses, ranging from classifying data to finding patterns in the same to improving employee relationships and managing customer preferences.

The application of AI, thus, should be done to add benefit to tasks that require huge effort but yield predictable results, so that the organizational leaders can further focus on finding ways to add value to the organization and leading them towards success through creativity and critical thinking. AI can add to the value by contributing towards doing things that it does best i.e. administrative things that are repetitive and require less cognitive abilities.

The computer revolution, development of microprocessors, automation, mechanization, rapid communication systems such as satellite communications, and electronics, have all revolutionized business. Despite this, it is extremely difficult to replace humans with robots, making human resources important. Human Resource management benefits a person in a variety of ways and at different levels. It assists employees in building teamwork and team spirit while also providing excellent career possibilities to those with the potential to advance.

Training and development inside an organization itself is a huge task that automation or AI implementation can take care of. An organization that wishes to expand continually needs human resources. However, in today's fast-paced economy, this is only achievable if organizations or businesses have proper personnel policies in place to inspire staff. Simultaneously, employees' skills must be constantly accessed, taught, developed, and enlarged. In an atmosphere of mutual trust, friendliness, and collaboration, they must be encouraged to take risks, experiment, create, and make things happen. A competent and motivated staff is necessary for any organization's existence and success. There has been extensive industrial strikes/unrest, trade union influence on the workforce, strained worker-management relations, and a rising divide between management and employees throughout the last few decades, among other things; and all of this is a wake-up call for integrating AI systems within organizations to pick up slack work that is important, given, but also does not necessarily require human attention.

An organization has many roles to play. It is a method for management in action because an organization is more than just a chart; all management actions such as direction, motivation, coordination, and control of business enterprises are clearly defined, reducing uncertainty and vagueness in employee performance. And an organization is nothing more than a group of employees in different roles and authority who no doubt bring knowledge, creativity, and moral judgment to the table, but our brains are also naturally programmed to make biased decisions in certain situations. Now, AI, on the other hand, uses facts and figures among data analysis and mechanical algorithms to counteract our biases. For example, for recruitment on senior levels, the implementation of AI can pave the way through a few simple biases which other executives may have in addition to using existing data to make decisions regarding hiring other non-executives, as well.

To implement AI in any organization it is very curtailed for the management and the leadership team to take an active role in deciding how should the capabilities of AI is to be used because AI has a wide set of applications, it can be used in production-related work or creating analytics using date to get in-depth insights or it can be used for improving customer experience, the applications are endless but the first step is to decide where and how AI can be utilized depending on the needs of the organization.

AI should be implemented in a company with a certain purpose and not just because it is the buzzword for marketing purposes; implementing AI without a purpose is a waste of tons of resources. Once the management team extends its support it is the responsibility of the mid-level managers to execute the work, the managers are responsible to coordinate with the tech team and work on feeding the AI with the required data.

It is critical to have a technical AI team in place as these individuals are the experts that will work first hand on the software and make necessary modifications as necessary. They will be responsible for taking managers' input to understand the end goals of the work they are doing and create meaningful insights for the management team making decisions, the nature of this team is highly technical.

Once the entire process is set up, it is required to keep the data safe and organized, it might seem like very simple work that may not require a lot of effort but the nature of data that we are dealing with is huge and it becomes extremely difficult to manage if it has not been organized from the beginning.

The size of the individual teams in the AI process depends on the size of the company and it is essential for any company willing to incorporate AI to have this process in place. Once the process and teams are in place, AI can be implemented on multiple fronts based on goals by replicating the process with a different purpose.

Let's understand this with an example: Assuming that an organization decides to use AI to improve its customer experience in the customer support department, the first step in the process is already done and the management has decided where they want to use AI. Now, the next step will be for managers to coordinate and get the data from the past; the data can be on general questions a customer asks when they contact customer support, data on solutions to those problems and data on what makes the customers happy, etc., Data is the foundation on which AI will work and the more data the better it is. The next step in the process will be completely technical and here the tech team will work on algorithms and design the program as per customer needs of the organization and probably create a chatbot that listens to customer problems and provides solutions without any human intervention, and later the data is organized by the data management people. Here AI is being used to improve the customer experience without human intervention, which will improve the efficiency and effectiveness of the company. This is just one example of how AI will be helpful for a company. This process can be replicated and mutated as per the purpose and finally help the company to move closer to its desired goal.

Here is another example of how a business can with the help of AI can improve their process or decision making. AI can be very efficiently used in the manufacturing of products by automating processes, where robots powered by AI with minimum human intervention manufacture the product; by using AI-powered machines the face of the entire assembly line will be modified.

AI has also been very helpful in the healthcare sector where it has been used on multiple fronts by reducing hospital visits of the patients via AI virtual assistants and assisting surgeons during surgery and it has also been helpful in the process of diagnostics and reduced misdiagnosis. AI has even changed the way we work in our day-to-day lives as our cell phones have AI-powered assistants like Siri or Google assistant that recognizes just by a vocal command and delivers the results, also customizing results to our preferences.

AI has a huge impact on how organizations work and in the strategic implementations of decisions and as the technology grows so do the possibilities of how it can be used. AI is a set of powerful algorithms and depends on the computational power of computers for its functioning efficiency and effectiveness; to say the very least, as the computational power grows so does the impact of AI. This is what brings us to quantum computing, let us understand what quantum computing is.

Quantum computing might be the future of computers. There are over 20 companies that are working on this technology to make it mainstream; the companies include tech giants like Google, IBM, Microsoft, and Intel. The technology is fairly new, and it is difficult to say when it will enter mainstream computing, probably another five years optimistically.

Quantum computers are powerful and work differently than traditional computers. Traditional computers process data in binary bits which can be either 0 or 1; but quantum computers are not like that, they process data in qubits which are particles like photons or electrons and represent 0 and 1 but it gets interesting that the data doesn't have to be processed in

0 or 1 but it can be both 0 and 1 at the same time and can be anywhere in between 0 and 1. In simple terms, by computing this way gives quantum computers an exponential advantage over regular computers.

It is essential to understand that quantum computers and classical computers use completely different technology, and it is not the case that we make classical computers better and better to build a quantum computer. For example, we can consider a bicycle and a motorbike although they both serve the same purpose of transportation, the way they work is completely different and it is not possible to build a motorbike by building a better and better bicycle.

Quantum computers can process huge amounts of data and solve complex problems in a matter of seconds, which would take classical computers thousands of years to solve.

As quantum computers are in their initial stages they have their own set of problems. The amount of heat generated is massive, and hence large cooling systems are needed and the entire system has to run in a controlled environment.

7.5 IMPACT OF QUANTUM COMPUTING ON AI

As we know, AI is essentially making a machine mimic human behavior but as of today's reality, AI can mimic human behavior for just a particular task and not entirely replicate a person, which is what we call narrow AI, meaning it is extremely good at performing one particular set of tasks. AI is a set of algorithms that are run on a computer and if these algorithms are paired with the computational power of a quantum computer it will exponentially increase the speed of the calculations, permutations, and combinations and will be able to process data in a matter of seconds. In simple terms, pairing quantum computers with AI is like running AI algorithms on performance enhancement drugs.

Having this powerful tool in hand, the possibilities of using quantum-powered AI in business become endless, as a large number of permutations and combinations can be simulated, and various patterns can be derived from the simulations which will make a fundamental impact on the way strategic decisions are made.

Quantum computers with their immense power merged with AI technology can create wonders in the healthcare industry by the discovery of various types of cures for diseases. With the computational power that we currently have, it is not possible to solve some complex problems but quantum computers can solve the problems and review molecules, proteins and chemicals by creating simulations and thus find a cure. Animal testing can also be reduced, not to mention the time and money saved that would be spent on research.

Quantum computers can also be used in security for encrypting data which will make it extremely challenging for anyone to break the code. As we know, these computers are great at creating simulations, and can be used in various cases like generating accurate weather reports and helping countries reach their goals to achieve sustainable development. They can be used in analytics for trading purposes and institutes can increase transactions and data speed and boost trade.

If we consider purely strategic decisions then various decision outcomes can be simulated with multiple variables and the result can be predicted even before the decisions are ever made. In addition, with the help of simulations, we can also find possible alternatives to reach a goal in the least amount of time.

The possibilities are endless, and more uses will arise as the technology develops. It is essential to understand that this great power of computing is a tool which can be used for the good and bad as well; with this kind of power in hand it is quite easy to surpass any type of cybersecurity system and even sensitive data can be accessed. Businesses and responsible bodies must take appropriate measures to counter and reduce the risks and use the technology for betterment.

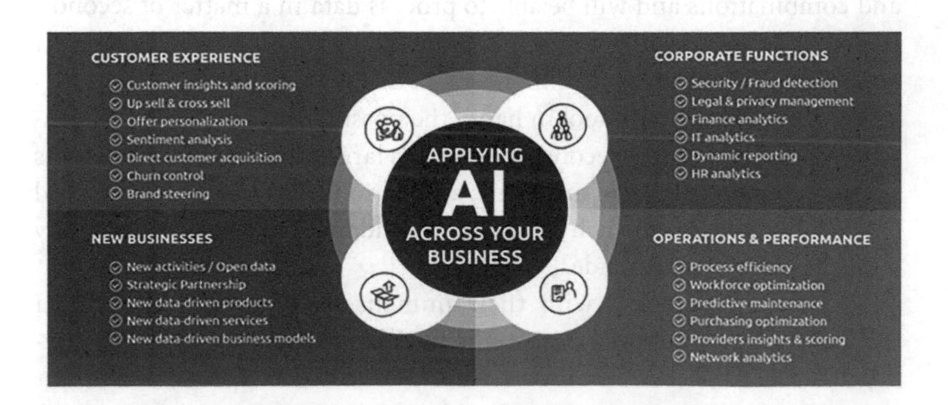

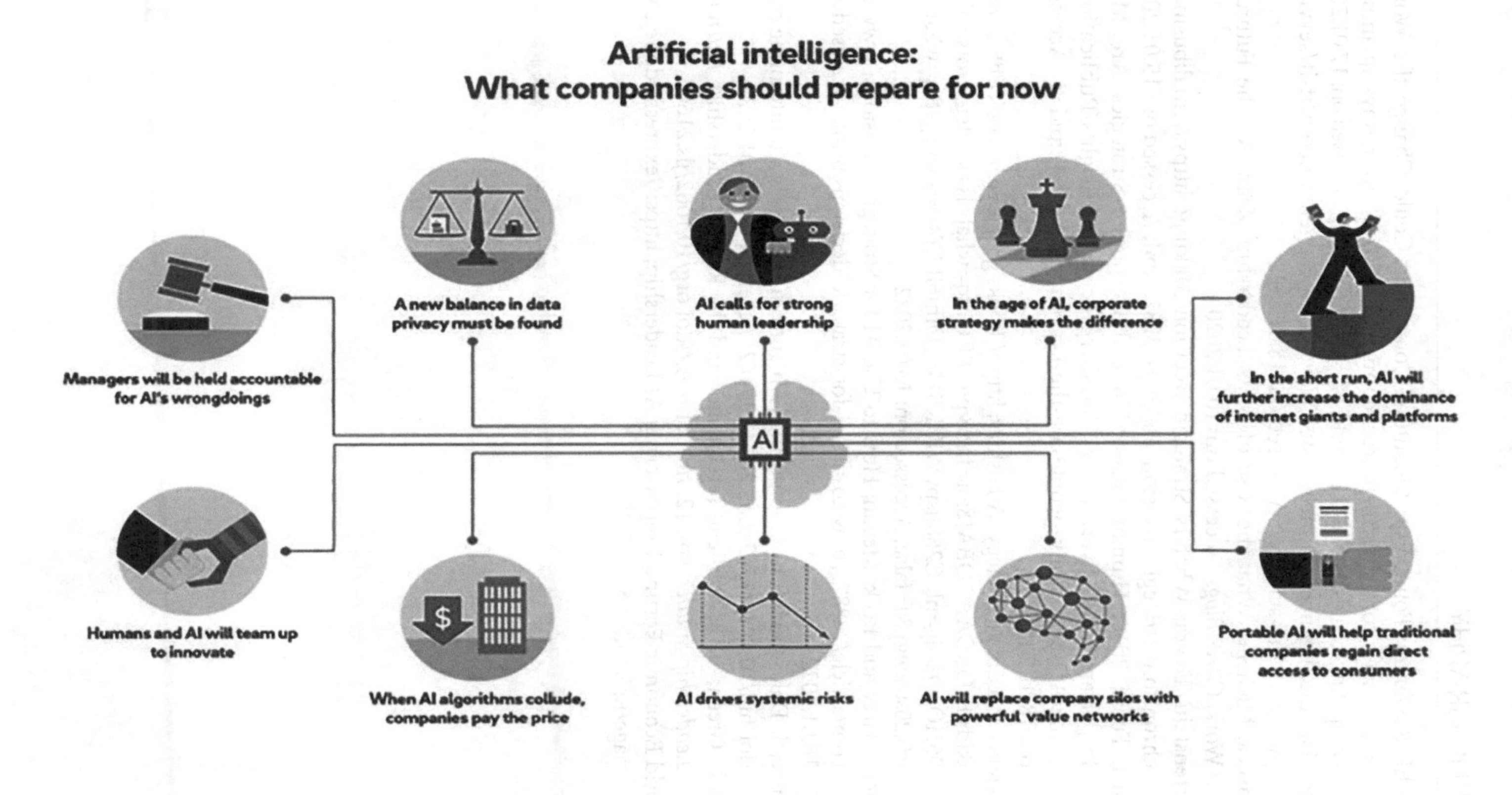
Artificial intelligence:
What companies should prepare for now
Managers will be held accountable for AI's wrongdoings
A new balance in data privacy must be found
AI calls for strong human leadership
In the age of AI, corporate strategy makes the difference
In the short run, AI will further increase the dominance of internet giants and platforms
AI
Humans and AI will team up to innovate
When AI algorithms collude, companies pay the price
AI drives systemic risks
AI will replace company silos with powerful value networks
Portable AI will help traditional companies regain direct access to consumers

BIBLIOGRAPHY

Ali EL Kaafarani; Four Ways Quantum Computing Could Change the World; www.forbes.com/sites/forbestechcouncil/2021/07/30/four-ways-quantum-computing-could-change-the-world/?sh=36de86eb4602; Accessed on: 17.01.2022

Artificial Intelligence: Pioneer the Way; www.capgemini.com/it-it/service/perform-ai/ai-activate/; Accessed on: 18.01.2022

Fonseca, Michelle; What the Rise of AI in Leadership Means for the Future of Work; *Citrix Blogs*; Accessed on: 10.01.2020

Gartenstein, Devra; What Is Strategic Decision Making? https://smallbusiness.chron.com/strategic-decision-making-23782.html; Accessed on: 15.01.2022

Klaus, Fuest; 2018; Human Leadership and Corporate Strategies Are More Important Than Ever; www.rolandberger.com/en/Insights/Publications/Using-AI-successfully-people-are-the-key.html#!#&gid=1&pid=1 Accessed on: 18.01.2022

Lawrence Freedman; Strategy: A History; https://books.google.co.in/books?hl=en&lr=&id=BeQRDAAAQBAJ&oi=fnd&pg=PP1&dq=what+is+strategy&ots=nhVs06HO5O&sig=nluC268uspvWiUK51W_Ofm6lLLY#v=onepage&q=what%20is%20strategy&f=false; Accessed on: 15.01.2022

Libert, Barry and Beck, Megan; How to Use AI for Strategic Decisions; www.cio.com/article/228877/how-to-use-ai-for-strategic-decisions.html; Accessed on: 16.01.2022

Prince, J. Dale; Quantum Computing: An Introduction; www.tandfonline.com/doi/full/10.1080/15424065.2014.939462; Accessed on: 16.01.2022

Smith, Green; 2018; Artificial Intelligence and the Role of Leadership; *Journal of Leadership Studies*, vol. 12, no. 3; https://doi.org/10.1002/jls.21605

World Economic Forum; Empowering AI Leadership; https://express.adobe.com/page/RsXNkZANwMLEf/

CHAPTER 8

The Synergy between AI, Quantum Management, Command and Control

Akash Gurrala

Eguturi Manjith Kumar Reddy

Juan R. Jaramillo

CONTENTS

8.1	Introduction	94
8.2	What Is AI?	94
8.3	What Is Quantum Management?	95
8.4	Command and Control	97
8.5	Advantages	98
8.6	Disadvantages	99
8.7	Synergy between AI, Quantum Management, Command and Control	99
References		102

DOI: 10.1201/9781003244660-8

8.1 INTRODUCTION

Organizations have set of rules, regulations, plan, teamwork and many other parameters in order to achieve their desired goal. For achieving goals, the resources should be structured in a way that they can be maximally utilized by an organization. Management, Command and Control methods play the most significant role in accomplishing the mission of an organization. They create a discipline within an organization and help them work efficiently in a properly structured way. Proper implementation of these methods enables an organization to survive in a changing environment. The traditional management methods such as the Newtonian method which is an extension of Newton's physics to be used in Management Science, are not able to perform better when it comes to solving today's complex problems. Here's where Quantum integrated with AI, Management, Command and Control comes into play and are being used together to solve complex problems in a most efficient manner. Further let's study AI, Quantum Management, Command, Control and the synergy between them in order to achieve a goal in an organization.

8.2 WHAT IS AI?

Artificial intelligence is a human-made intelligence which is demonstrated by machines or robots to perform and execute decision making. The way of processing is similar to that of humans and sometimes more efficient. AI plays an important role in Assistance and Automation. We have some existing artificial human assistance platforms like Alexa, Google Assistant and Siri which can receive commands from the user, understand them and thereby execute the task accordingly. Operating Systems with embedded AI will automatically identify problems and rectify them. This assistance comes into play in three contexts,

I) They execute the commands they are given.

II) They help in completing the tasks in parallel with masters.

III) They proceed with their own designs.

When we ask Siri to call someone, she will do it. This example comes under context I where there is less human effort, this is day-to-day assistance we generally experience. In context II, humans work with the support of AI, when complexity increases in making decisions, we humans need computational support to succeed, and also, AI needs the help of humans to make decisions which involve emotional intelligence. In context II, human intelligence and AI depend on each other. Advanced AIs such as Tesla come under context III. These kinds of AIs are mostly used in automation and decision making independent of human intelligence. Spacecraft, satellites and rovers such as *Curiosity* which have some specific tasks are improvised or changed according to the circumstances. The rovers or satellites adapt to space or any other planet's habitat and execute their jobs, which are achieved by complex AI algorithms.

When any particular research data collected by a human researcher makes them drown in data and show difficulties in managing the data. Whereas AI applications which make use of machine learning methods can intake the data in less time and can convert it into useable/actionable data [1].

8.3 WHAT IS QUANTUM MANAGEMENT?

After the expansion of Quantum Theory in the field of biology, chemistry, etc., Quantum Theory has even extended to psychology, intelligence and consciousness which ultimately started gaining the interest of management scholars and entrepreneurs entering the field of management science, and thereby established a management of new methodology termed 'Quantum Management'. Quantum Theory and Management Science are synergized to be called Quantum Management which is a derived topic from Quantum Theory.

In 1900, the German physicist Planck proposed an important concept named 'Quantum' to explain experimental phenomena in blackbody radiation. In 1905, Albert Einstein evolved the quantum theory and explained the light quantum hypothesis. In 1913, Bohr's atomic model was constructed by applying the quantum theory to the hydrogen atomic system. Broglie proposed the wave-particle duality nature of matter in 1924. Since then, the quantum concept is being used to explain the fundamental laws of the microscopic material world and applied in chemistry, and other fields. Based on the practical studies on the bridge between quantum theory and human thoughts conducted by some scholars, using Kolmogorov's probability theory or using a geometric framework, quantum theory has expanded to explain the differences in the human process of acquiring knowledge and decision making [2].

Quantum Management is among the management techniques that acknowledges the human systems like companies function best when led, managed and structured to work like natural systems. As said earlier, the main aim of quantum management is to boost the ability and efficiency of employees and managers making the organization to be more productive. The quantum management model appears to be comprehensive and structured in the present and the future. Through communication and conferences quantum managers can bring out solution for the problems of respective companies. When compared to traditional management methods, the attributes, skills and jobs of managers are different in quantum management, as quantum managers manage the organization and Human Resources thoughtfully and rationally and also try to manage the equilibrium between stress and discipline. The execution of quantum ways helps for finer explanation of the distinction between human judgement and choice than that of traditional management systems.

Characteristics of Quantum Management:

i. *Integrity*: Enterprises are correlated and cooperatively described in the global environment.

ii. *Flexible*: This method of a management system is said to be uncertain and complex. It is between existence and potential, particle and wave, etc., which indicates it must be flexible.

iii. *Self-Organizing*: Quantum Management must be carried out by structural innovation using discrete creative facilities.

iv. *Plurality*: This management system has numerous ways to adjust to the diversity of a community, market and each fellow.

v. *Improvisation*: To improvise the time-to-time performance of the whole organization.

vi. *Fun and value seeking*: Supports play and reward, values mediations, goals and life morals.

vii. *Participative*: Quantum management has transformed the environment between humankind to create its own life, well connected to the surroundings [2].

TABLE 8.1 Newtonian Physics and Taylorian Company vs Quantum Physics and Quantum Company

Newtonian	Taylorian	Quantum	Quantum Company
Mechanical	Company: A Machine	Adaptive System	Conscious Living System
Atomistic	Siloed Functions	Holistic; Entangled	Network Organizations
Deterministic	Bureaucratic; Rule-Bound	Spontaneous; Self-Organizing	Self-Organizing
Particle or Wave	Particle-Like; Division of Labor	Both Particle and Wave	Employees: Individuals and Team members
One Best Way	One Best Way; Simple Point of View	Superposition of Multiple Potentialities	Many Points of View
React to Forces	Reactive	In Dialogue with Environment	Responsive; Agile; Adaptive
Observers Passive Witnesses	Employees Passive; Units of Production	Participatory Universe	Employees Co-Creative Partners
Isolated	Company Isolated from Environment	Contextual	In Dialogue with Environment and Customers; An Ecosystem
Absolute	Leader Knows Best; No Questions	Heisenberg; Questions Determine Answer	Encourages Questions

Table 8.1 shows the comparison between Newtonian physics and Taylorian company vs Quantum physics and Quantum company [3].

8.4 COMMAND AND CONTROL

Command and control is a management approach to leadership with strict authority and formal instructions [4]. In order achieve a goal or accomplish a mission we need clear plan, great thought process as well as a proper vision, it results not only in achieving the goal but also helps the organization maintain its success. 'Command and control' is the implementation of exertion of authority and direction by a properly designated commander over assigned and affiliate forces in the achievement of the target. Command and control functions are performed through an arrangement of personnel, equipment, communication, facilities, and procedures employed by a commander in planning, directing, coordinating, and controlling forces and operations in the accomplishments of the mission [5].

Command is the broadcast of a vision to the organization to better help accomplishing a mission. Command intensifies success and reward. That is, in order to survive the organization has to be successful and encourage by both intrinsically and extrinsically rewarding its members. Control is the process concerned to deal with uncertainties by establishing and providing structure. The changes produced by vision are likely to also produce tension. The leaders of the organization must deal with these uncertainties so that they do not affect the organization [6].

Organizations that are characterized by command and control generally:

- Have a centralized decision-making system.
- Only allow senior management to set the rules.
- With an increase in seniority, salary perks and flexibility also increase.
- The higher you go the more information is privatized.
- Have specialized internal departments such as Human Resources.
- There is a large distinction between upper management and the employees.
- Possess a disciplined culture that fills orders quickly [4].

Command presents the vision/mission to the people who can best implement it while control is concerned with reducing risk and making the process more efficient.

8.5 ADVANTAGES

- *Responsibility*: Every employee in an organization has individual responsibility and their supervisors are supposed to handle them.
- *Productivity*: The management method here helps them to maintain a good workflow which results in decent productivity.

- *Good relationships internally and externally*: Good relationships not only allow you to maintain good employee relationships but also helps the organization to build strong relationships with stakeholders, vendors and customers.
- *Boost employee morale*: As employees are allowed to share their concerns with supervisors, the problems employees face are resolved which in turn boosts their morale [7].

8.6 DISADVANTAGES

- *Less collaboration*: Only upper management sets the rules. There may be a case where the middle managers and the employees have autonomy in their work which might result in less collaboration between higher authority and employees.
- *Reduced employee empowerment*: Since the employee may not be able to make a decision in a particular situation, it can be discouraging because employees can't act on their own in the respective situation.
- *More competition*: Increase in competition may lead to a factor of mistrust among peer managers.
- *Slow communication*: Since an idea or issue has to go through various stages in order to reach the decision maker, this may slow down communication. Suppose an employee has to point out a problem and they expect an action from the organization. The employee explains the problem to their supervisor, then the supervisor communicates with the middle manager and the middle manager may decide to talk or ask for advice from their senior manager, which ultimately results in slow communication.

8.7 SYNERGY BETWEEN AI, QUANTUM MANAGEMENT, COMMAND AND CONTROL

Artificial Intelligence and Quantum Computing are two technologies that are considered to have significant benefits in the industrial evolution. AI most likely requires Quantum Computing to achieve better progress when compared to traditional computational methods. This also helps to solve complex problems with more efficiency and accuracy. The synergy between

Quantum Computing and Artificial Intelligence is termed 'Quantum AI' which is currently evolving technology that makes use of quantum computing for computational power for machine learning algorithms [8]. Since Quantum AI can also be used for prediction purposes, it helps an organization to know what they are lacking/what they need to better function to reach their goal. Also, one of the attributes of quantum management is being adaptive, i.e., without any huge effort the management system can easily adjust to the respective situation and respond accordingly, the organization can easily change its structure and plan for better management and thereby achieve their goals.

Having multiple states of success instead of just one. An organization following traditional methods defines success as just one state and does vigorous work to achieve it. Whereas using AI, Quantum, Command and Control together defines the success in multiple states. For example, let's say revenue is a state of success for an organization, quantum methods also consider other parameters such as a large number of people using the product, market value, etc., to be multiple states of success. As the greater number of states of success are defined, the more opportunities the organization will see which in turn results in more milestones and merits. The beauty of this synergy exists in discovering new techniques for collecting data about the unobservable. Not everything can be easily observed/extracted from the environment, that's where the synergy comes into the play and performs better in gathering information from unobservable/barely visible environments, which helps to extract the hidden patterns from the progress, noticing drawbacks, solving drawbacks and increasing in efficiency of performance. This synergy can also be used to motivate and entangle the team, which in turn results in increased acceleration of progress, quality and also to boost ethics in an organization. A few methods of entanglement are listed below:

- *Accountability*: Encouraging employees by holding yourself to a set of standards and discipline. Members can be inspired by leaders. So, once the manager has accountability, i.e., maintaining a set of standards, it's most likely the employees will also mirror their accountability. The more the leader is accountable, the more accountable the team is. This method is an approach of encouraging the team to be clear about what they have to do and what they are going to do by behaving in that way first.

- *Empathy*: Trying to empathize with the customer by knowing the product that they truly need. When we observe ourselves as a user, then we can easily understand customer requirements and based on that working on the related product. The entanglement with the customer helps us go through how their experience is on a deeper level and what improvements can be made.
- *Incentive*: Bringing out better results by linking the team's reward to the success of the project. This helps in motivating employees work much better and also provides encouragement to work on many other projects too. The rewards need not always be financial but instead they can also be visibility within the organization, career promotion, etc., which help the employee not only to perform better but also creates a subconscious passion for the work done and extends their potential. When the employee is truly working hard, they should be noticed for their hard work and receive any form of appreciation. Not getting these types of reinforcement can limit employee potential, and employees can lose interest, etc.
- *Togetherness*: Building a healthy environment among peers initiates better collaboration, boosting creativity, increasing productivity, reducing stress, etc. The manager can take their team to dinner helping them personally, making the work environment more free for the employees as well as managers. This connection encourages the people of an organization to know each other on a personal level.

The implementation of AI, Quantum Management, Command and Control together might make a manager difficult to recognize if their quantum leadership method is being used effectively because we exist in both a state of success as well as failure concurrently. Looking for constant feedback from people outside of this management may allow us to grow, stretch and be more flexible besides limiting our own possibilities. There's also a mistake senior management makes in traditional management methods, thinking they know and can do it best. It's a responsibility of the manager to look after their team by not only assuming that they have the best solution for the respective problem but rather knowing the whole team's approach towards the problem and encouraging the individual ideas too. There might be a case where the manager's ideas were best and also there might be a case where employees/whole teams came up with an idea that is more efficient. This way the manager can also perform efficiently by bringing out the best from their team [9]. The usage of Quantum AI for an organization may be the best for adapting, problem solving, progress analysis, self-observance, constant improvements, better communication, better predictions, implementation of entanglements and for many other purposes. This way an organization can succeed inside and outside by implementing the synergy between AI, Quantum Management, Command, and Control.

REFERENCES

[1] E. Burns, N. Laskowski, and L. Tucci, "What Is Artificial Intelligence (AI) and How Does It Work?—Definition" *TechTarget*. www.techtarget.com/searchenterpriseai/definition/AI-Artificial-Intelligence (accessed Jan. 13, 2022).
[2] X. Yin, "Review and Prospect of Quantum Management," *Am. J. Ind. Bus. Manag.*, vol. 9, no. 12, pp. 2220–2230, 2019, doi: 10.4236/ajibm.2019.912147.
[3] D. Zohar, "What Is Quantum Management?" *Zero Distance*, pp. 41–53, 2022, doi: 10.1007/978-981-16-7849-3_4.
[4] John Spacey, "7 Characteristics of Command and Control" *Simplicable*, May 5, 2020. https://simplicable.com/en/command-and-control (accessed Jan. 11, 2022).
[5] "Command and Control—Glossary" CSRC, https://csrc.nist.gov/glossary/term/command_and_control (accessed Jan. 10, 2022).
[6] "Leadership, Management, Command, and Control," www.nwlink.com/~donclark/leader/LMCC.html (accessed Jan. 11, 2022).
[7] Williams Roman, "Chain of Command: Definition, Advantages, Disadvantages," *Small Business Journals*, Apr. 29, 2021. https://smallbusinessjournals.com/chain-of-command/ (accessed Jan. 12, 2022).
[8] C. Dilmegani, "In-Depth Guide to Quantum Artificial Intelligence," *AIMultiple*, 2020. https://research.aimultiple.com/quantum-ai/ (accessed Jan. 8, 2022).
[9] J. Everingham, "The Principles of Quantum Team Management" *First Round Review*. https://review.firstround.com/the-principles-of-quantum-team-management (accessed Jan. 13, 2022).

CHAPTER 9

Quantum Impact on Organizational Performance

Preetham Reddy Gandagari
Debdutta Choudhury
Daya Shankar
Jorge A. Wise

CONTENTS

9.1 Introduction 104
9.1.1 What Is Quantum Computing? 104
9.1.2 "0 or 1" vs "0 to 1"; Bit vs Qubit 105
9.1.3 Where Are We on Quantum Computers? 105
9.1.4 Applications 105
9.2 Q2B (Quantum to Business) 106
9.2.1 AI with Quantum Enhancements 108
9.2.2 Implementation of Quantum in Businesses 108
9.2.3 Select 108
9.2.4 Identify 108
9.2.5 Outline 109
9.2.6 Adapt 109

DOI: 10.1201/9781003244660-9

9.3 Quantum Computing in Practice 109
9.3.1 Sycamore 109
9.3.2 Finance (Banking) 109
9.3.2.1 Partnerships and Collaborations 110
9.3.3 Chemical Industry 110
9.3.4 Cybersecurity 111
9.3.5 Bio-Medical and Pharmaceutical Companies 112
9.3.6 Government 113
9.3.7 Military 113
9.4 Quantum Leap Ahead 114
9.5 Conclusion 115
Bibliography 116

9.1 INTRODUCTION

Quantum technology has the potential to become a disruptive field (Benioff, 1982). It was noted that a quantum mechanical model of a microscopic level can be constructed and such models should be examined, as it was observed that these machines have advantages over the classical large computers built in 1970s. However, it was noted that the said model was very complex and was stated that it is difficult to construct this kind of machine based on this model. This was the first paper of its kind to suggest that computers can be built using quantum mechanics.

If we want to make computers the size of atoms which are extremely small we cannot do so using the classical mechanics instead we need to use the laws of quantum mechanics. This is where the future of computations—Quantum Computers arrive (Feynman, 1982). Richard Feynman was the first to coin the term Quantum Computer and reasoned that classical computers cannot simulate the physical phenomenon that a quantum computer can simulate.

9.1.1 What Is Quantum Computing?

Classical computers are based on classical mechanics which primarily deals with the motion of particles that are visible to the eye. This theory does not apply to atomic size particles and less than atomic sizes i.e. subatomic particles, which is where quantum mechanics enters the picture. Quantum mechanics deals with the motion of subatomic particles (Brassard et al., 1998).

9.1.2 "0 or 1" vs "0 to 1"; Bit vs Qubit

While classical computers also known as binary computers store data in bits which are represented by either 0 or 1, that is, either "on" or "off". Which means only one of the two states are stored in the bits. When it comes to a quantum computer, a bit is called a Qubit and can be either 0 or 1 or "any combination of the 2 states" at the same time; which is called a superposition. This is how quantum computers cut short the time to perform complex calculations. When two superpositions are connected or stay entangled or synchronized and if you perform an operation on one qubit it will have an instantaneous effect on the other qubit, this is called *quantum entanglement* (Rieffel and Polak, 2000).

9.1.3 Where Are We on Quantum Computers?

Quantum computers are categorised into three types in the order or a path from a limited quantum computer to a potential full-fledged quantum computer. The first type is quantum annealing; which are limited and these are not on a path to quantum computer as they cannot offer a reasonable speed bump over current classical computers.

The second is NISQ (Noisy Intermediate Scale Quantum) computing, these have error correction (detecting errors and reconstructing the data that is error-free). These computers are best for business applications and are on a path towards the full-fledged quantum computer which is also the third type called Fault Tolerant Universal Quantum computing. These are capable of handling business as well as scientific calculations faster than classical computers (Gil et al., 2018).

Currently, we are in the era of NISQ computing, qubit system errors have to be rectified, i.e. they might lose information in the presence of noise. This stage will last anywhere between two to nine years and we can still reap the benefits of designing new chemicals, optimizing investment portfolio and also used in the pharmaceutical industry in discovering drugs.

9.1.4 Applications

Quantum computer applications are applicable in almost all aspects, however, it is not recommended to seek quantum computing for small tasks which can be easily done by the classical computer. Quantum computing is used for extremely complex tasks involved in designing new chemicals, molecular structures, all kinds of simulations, predictions, drug discoveries, portfolio management, optimisation techniques, machine learning, artificial intelligence, route optimizations, search engines, research, encrypting and decrypting, etc.

Because general-purpose quantum computing will almost certainly never be cost-effective, quantum computing applications will be specialised and focused. However, the technology has the potential to transform a number of industries. Quantum computing has the potential to enable advances in the following areas:

Machine Learning: Improved machine learning through faster structured prediction. Boltzmann machines, quantum Boltzmann machines, semi-supervised learning, unsupervised learning, and deep learning are only a few examples.

Chemical Industry: New fertilisers, catalysts, and battery chemistry will all help to increase resource usage.

Pharmaceuticals: New medications, personalised drugs, and possibly even a hair restorer are all on the horizon.

Finance Organizations: Quantum computing could make Monte Carlo simulations faster and more complicated, allowing for things like trading, trajectory optimization, market instability, pricing optimization, and hedging methods.

Healthcare Sector: DNA gene sequencing, such as optimising radiation treatment or detecting brain tumours, might be done in seconds rather than hours or weeks.

Metallurgy: Super strong materials, corrosion-resistant coatings, lubricants, and semiconductors are some examples.

Computer Science: Faster multidimensional search functions in computer science, such as query optimization, mathematics, and simulations.

9.2 Q2B (QUANTUM TO BUSINESS)

The important aspect is to select the right quantum computer depending on the business. Organizations that are paying attention to quantum computing and implementing it as early as possible will snatch the leadership position in the market from their competitors.

Commercialisation of quantum computing will bring about benefits in mainly three quantum-assisted areas namely: simulations, optimisations, and machine learning.

Simulation of quantum mechanics: Quantum computing can mimic naturally occurring processes and systems because quantum mechanics is fundamentally linked to natural phenomenon. This powerful capability could pave the way for electric vehicle manufacturers to produce longer-lasting batteries.

- Drugs customised for an individual patient could be developed quickly by biotech businesses. Electric power transmission costs could be decreased. Fertilisers might be made more effective, with significant implications for food production around the world.

Optimization at the quantum level: In a scenario when several viable solutions exist, the art of addressing optimization problems entails identifying the best or "optimal" one. Consider the task of creating a parcel delivery schedule.

- Mathematically, there are more than 3.6 million different ways to schedule ten deliveries in close proximity. But, given factors such as recipient timing constraints, probable delays, and the shelf life of delivered commodities, we cannot ascertain which could be the best option.
- Even with approximation techniques, the number of possibilities is considerably too huge for a traditional computer to investigate. As a result, traditional computers nowadays use a number of shortcuts to solve large optimization problems. However, their solutions are frequently inadequate.

Benefits for businesses from using quantum computing: In order to stay ahead of the competition, businesses are already experimenting with quantum computing to find optimal solutions. After the first demonstrations of quantum advantages in optimization are confirmed, their prediction may deliver significant benefits.

- Universities scheduling classes are among the businesses that could benefit from quantum optimization
- Improving network efficiency for telecommunication companies

- Healthcare providers optimising patient treatments
- Air traffic control improvements
- Personalised offers from consumer goods and retailers
- Improving risk management services for finance companies
- Work schedules being developed by businesses

9.2.1 AI with Quantum Enhancements

Quantum computing has the potential to expand AI's capabilities by allowing it to explore a large number of options that a traditional computer cannot. In reality, a virtuous cycle of growth in both domains is beginning to emerge as a result of a synergy between AI and quantum computing.

Quantum algorithms, for example, can help machine learning improve data clustering, while machine learning can help us better comprehend quantum systems.

Quantum-enabled cognitive computers could one day infect almost every industry, giving professionals with advanced, proactive decision support, employees with targeted, responsive training, and customers with specially personalised, adaptable vendor relationships.

9.2.2 Implementation of Quantum in Businesses

Organizations that implement NISQ soon will be able to outgrow their competitors so much so that they will be able to optimise on a level that has never done before to invent new products. A path for quantum supremacy can be achieved by the following steps.

9.2.3 Select

Selecting some of the professionals from the company to educate them in quantum computing or whose expertise lies in this area. They should be tasked with understanding how quantum computing benefits the industry you are in and ask them to report with how your industry peers are responding and how might you Organization benefit from using quantum computing.

9.2.4 Identify

Any firm wishing to adopt quantum computing technology needs to identify the business processes that can be pivoted using quantum computing. Identification, tracking, and development of quantum-based applications are mandatory to generate long-term competitive advantages. Some of the potential areas are in the domain of innovation, product development and supply chains, which have huge commercial implications.

9.2.5 Outline

A firm needs to create a quantum computing roadmap, including possible future stages, with the goal of pursuing issues that could pose strong competitive hurdles and provide a long-term competitive advantage. The important members of the firm can join a quantum community to help their company become quantum ready faster. This can assist in gaining greater access to technical infrastructure, industry-changing applications, and other resources.

9.2.6 Adapt

Quantum computing is advancing at a rapid pace and therefore looking for technologies and development toolkits that are quickly becoming industry standards and forming ecosystems. Recognizing that fresh advances may force one to rethink the quantum development strategy, including switching ecosystem partners. It has to be kept in mind that own quantum computing requirements may change over time, especially as one gains a better knowledge of which business problems can benefit the most from quantum computing solutions.

9.3 QUANTUM COMPUTING IN PRACTICE

9.3.1 Sycamore

The Sycamore quantum computer containing 53 qubits was used to perform a complex operation in 200 seconds and the same calculation on a standard computer would take at least 10,000 years to compute. It ran one million times to measure the output strings by sampling the circuit to calculate the probability distribution.

However, it must not be misconstrued to think that quantum computers are fully operational and they have come to commercialization. This is one of the steps to quantum success (Arute et al., 2019).

9.3.2 Finance (Banking)

The application of quantum computing in finance is on the rise due to the large processing ability. Simulation and portfolio management are the two most promising applications.

Vijayalakshmi, 2019 insisted, more than two-thirds of equity investment's have been realized since 2018. That's over $1.3 billion. Around 75% of investments are made on hardware and the rest on software. Here we can understand how important hardware is as the physical possession of an asset that can be patented and tough to imitate is a game changer for any company and the investor who invests in the said company.

9.3.2.1 Partnerships and Collaborations

- J.P. Morgan Chase and Barclays Bank have been contemplating quantum computing and started experimenting with it in 2017. J.P. Morgan has partnered with IBM to investigate the use of this technology in financial services, giving them direct access to IBM Q cloud-based tools for conducting business trials.
- Trading strategies, optimising portfolio, pricing of assets and analysing risk are currently areas of interest for J.P. Morgan Chase, which are being pursued through the use of quantum computing, which necessitates the employment of complicated mathematical models.
- MUFG and Mizuho, two Japanese banks, have teamed up with IBM Q Hub at Keio University to test future quantum computing applications in the financial sector. With Tohoku University in Sendai, Japan's Nomura has initiated a joint research effort on employing quantum computing in asset management.
- CBA has partnered with Telstra, the Federal Government, the State Government of New South Wales, and the University of New South Wales (UNSW) to launch Australia's first quantum computing enterprise.
- NatWest, a British retail bank, has teamed up with Fujitsu on an initiative to improve its mix of liquid assets, such as bonds, cash, and g-secs. NatWest bank's quantum technology team has completed extremely complicated computations on the bank's £120 billion liquid assets holdings at over 300 times the speed of a classical computer-based cloud mainframe (Budde and Volz, 2019).

9.3.3 Chemical Industry

The chemical industry is assured to be the initial beneficiary of quantum computing abilities. For the chemical industry, quantum computing is going to open a wide view of abilities to modelling subatomic particles at precision. This would result in discovering or inventing new molecular structures which are effective to complete different tasks and all these before even touching or creating any molecules in the lab through simulations. This kind of computational power could exponentially increase R&D growth in developing new products that were previously never possible to create.

This kind of computational power will tip the scales between companies in the chemical industry as those who adopt it early will reap the benefits and the others will be unable to catch up to those who do implement the resources of quantum computing (Kühn et al., 2019).

There are three steps recommended to chemical companies in launching quantum computing activities.

1. Companies need to figure out the opportunities provided by the quantum computers. They need quantum service providers with experts in the chemical domain to better understand the applications of the new quantum technology in the chemical domain.
2. In order to succeed you need a strategy so that you can develop solutions along with the emerging or undiscovered start-ups. Which can result in co-developing and collaboration with such companies as chemical companies might not have access to their own hardware, they need to know how to secure access to a quantum service provider.
3. Chemical companies also should consider building dedicated departments to identify, implement, and test use cases throughout the organization. These departments need to be attractive to new talent in quantum computing to secure the future.

Finally, quantum computing will be able to enable these chemical companies which try to implement the above steps to lower their costs and time and which in turn add to their net profit. Quantum computing is going to take the chemical industry by storm, the question is should they choose to accept the benefits it will bring to the table (Budde and Volz, 2019).

9.3.4 Cybersecurity

The use of quantum technology in cybersecurity has pushed government attention and public sector investment more than any other application. "Quantum computing could render today's cybersecurity outdated,"

World Economic Forum (WEF). When the United States enacted the Quantum Initiative Act in 2018 to fund quantum computing, cybersecurity became a hot topic.

The knowledge that the technology behind current cryptography uses combinatorics underpins this focus and investment. With a sufficiently big and coherent quantum computer, Peter Shor demonstrated in 1994 that some types of encryption will become far less difficult to break.

With a sufficiently big and coherent quantum computer, Peter Shor demonstrated in 1994 that some types of encryption will become significantly easier to break. To put it another way, a commonly used encryption standard would be jeopardised.

While quantum algorithms are applicable to a wide range of combinatorics calculations, cryptography is a straightforward and direct application. Shor's method especially brings to light the possibility for tackling large-scale problems with completely operational and efficient quantum computers. That isn't to say that today's security systems won't fail in the near future (Denning, 2019).

Fear of a Y2K bug led to significant computer system upgrades, and fear of a possible quantum computer employing Shor's algorithm means that constructing quantum-safe encryption solutions will be smart in certain areas. It's worth noting that knowing the limitations of quantum computers are crucial at this point. Quantum computing's potential should result in a shift in cryptography techniques, such as greater key lengths, but not the end of encryption (Bova et al., 2021).

9.3.5 Bio-Medical and Pharmaceutical Companies

Enzymes have always piqued the curiosity of pharmaceutical businesses. These proteins facilitate a wide range of biological reactions, frequently by precisely targeting a specific type of molecule. Using enzymes to their full potential could aid in the treatment of today's main ailments.

The molecular structure of enzymes is baffling and these very complicated structures are tough to simulate using classical computers. A powerful quantum computer, on the other hand, might properly forecast the characteristics, form, and sensitivity of such molecules in a matter of hours, a breakthrough that could transform medication research and lead to a new age in healthcare.

Quantum computers have the ability to solve problems of this size and sophistication in a wide variety of industrial applications, including banking, transportation industry, chemical industry, and cybersecurity, to name a few. Quantum computing, a vastly different approach to computing, aims to solve the unattainable in a few hours of processing time, uncover answers to problems that have bedevilled science and society for years, and open enormous potential for Organizations of all types.

Several companies will not realise significant benefits from quantum computing for at least a decade, with only a few seeing gains in the next four years. However, because the possibility is so tremendous and technical advancements are occurring at such a quick pace, every business leader should have a fundamental knowledge of how system functions, the types of problems it can assist in solving, and how to prepare to tap into its possibilities (Ménard et al., 2020).

9.3.6 Government

The President of the United States signed a National Quantum Initiative Act (NQIA) in 2018, which sanctioned over $1 billion for increasing investments in Quantum Technologies and also developed a "Quantum-Smart" workforce. It also established a National Quantum Coordination Office to advise the White House on all things quantum.

Finland's Ministry of Economic Affairs has funded VTT Technical Research Centre, which is owned and controlled by the government of Finland to build its own quantum computer.

Finland has been using Quantum technology since 1980s in the form of essential components for aerospace, brain imaging systems and terahertz imaging—industrial and medical imaging (Stantchev et al., 2020).

India's Ministry of Electronics and Information Technology launched "Q-sim," a Quantum Computer Simulation Toolkit to provide a first of its kind QDE (quantum development environment) in India to students, professionals, academicians and the scientific community in India at a budget of over $1 billion.

9.3.7 Military

Earlier, quantum technology has brought a revolution in the fields of imaging, nuclear, lasers (currently laser weapons being tested), semiconductors. Quantum Warfare is a term coined for the usage of quantum technology in all aspects of warfare, for example, nuclear weapons, laser weapons, etc.

Similarly, a term called Quantum Attack is used when someone uses quantum technologies to break into servers or any personal space for any purposes, be it security, eavesdropping, reconnaissance, etc. Like the Remote Sensing the Quantum Sensing will be used in magnetometers that use quantum technologies, quantum radar, etc. The quantum technologies used in the military need to effective, precise and have unique capabilities (Krelina, 2021).

There are a few terms that are being used in the military based on the usage of quantum technologies to either build new devices or significantly improve current technologies using quantum power.

Quantum Radar: This works like a traditional radar, and with the usage of quantum we invite precision into its advantages. However, the Q-radar is sensitive to noise as we are still in the NISQ era hence, a hybrid Q-radar is under development sacrificing on the sensitivity, but more than the current radar systems.

Quantum Ghost Imaging: This allows for detecting objects which are not in our line of sight. The image is created from the impending correlation that is generated through entangled photons.

Quantum Illumination: Similar to Ghost Imaging, this also uses two photons, however, one is kept idle and the other is sent and reflected from the target and the correlations are measured. Advantages of such devices is that they are impervious to noise.

Quantum I-S-T-A-R: Intelligence, Surveillance, Target Acquisition, and Reconnaissance; it is all in all operations for a wide situational awareness majorly impacting data gathering, processing, and target identification.

These are just the tip of the Quantum-berg, there are more applications that are currently being developed for Quantum Space warfare, underwater warfare, electronic warfare, biological and chemical simulations, and detection and new material designs, etc. (Krelina, 2021).

9.4 QUANTUM LEAP AHEAD

Clearly, the finance industry is going to reap the ultimate benefits from quantum computing be it in the short term or for the long haul. Then the global energy and materials is where the chemical industry steps in as there are going to advancements in inventing new materials and discovering new compounds, structures and more.

Advanced industries which include aerospace, defence, ships and submarine building, semiconductors are all going to have a good run in the medium and long term through the usage of quantum technologies. Pharmaceutical and medical industries will see increased usage in gene sequencing, drug discovery, vaccine discovery, medical imaging and various other aspects, this particular industry will have a significant positive effect in the long run.

The rest of the industries will be affected by advancements in the other industries but not that significantly as there are a lot of factors that need to be considered, however, these are the predictions based on quantum computing use cases as of 2019 (Ménard et al., 2020).

9.5 CONCLUSION

There is no doubt that qubits are going to play a crucial role in the future of computing. The fact that qubits open up a new world of computing that is so fast that the classical computers will not be able to catch up in the future. The physical states of quantum superposition and quantum entanglement offer significant computational advantages in every aspect of electronics. Quantum advantage will be able to solve many computational complex problems in the fields of finance, pharma, bio-medical, military, advanced industries, healthcare sector, IT, energy sector, chemical industry, etc. There is an incredible advancement in the manufacturing of hardware related to quantum development and quantum software by designing algorithms and all these have made quantum computing closer to reality.

We must also keep in mind that with all the advancements there are still some challenges that need to be resolved. For example, in the case of a quantum computer's chip, the cells are needed to be maintained at a very low temperature and with the slightest increase in temperature the cells could lose the information. Hence, the chip needs to be super cooled and maintained in that state. This does not have to be a deal breaker, as the current computers were in the size of a room with only few MBs of data. Only after years of development do we have these powerful devices in our hands. However, we do have to wait longer for a quantum computer to be commercialized even with all the advancements there are a lot of factors that need to be accounted for before public use.

Even though with the use of quantum computers on our current encryptions could decrypt it very easily, fear not, as there are developments in the field of Hybrid Quantum technology. Where we get the best of both worlds, the protection from a quantum computer with all the other aspects being the same until quantum computers are commercialized.

The question still remains, with all that computing power, what usage or benefit does it have to a common user who does not need high computational power.

BIBLIOGRAPHY

Andreas, B., Guillaume, B., Binder, J., Thierry, B., Ehm, H., Ehmer, T., . . . Winter, F. (2021). Industry quantum computing applications. *EPJ Quantum Technology*, *8*(1).

Arute, F., Arya, K., Babbush, R., Bacon, D., Bardin, J. C., Barends, R., . . . Martinis, J. M. (2019). Quantum supremacy using a programmable superconducting processor. *Nature*, *574*(7779), 505–510.

BCG Linkedin. Available from: www.linkedin.com/company/boston-consulting-group/.

Benioff, P. (1982). Quantum mechanical Hamiltonian models of Turing machines. *Journal of Statistical Physics*, *29*(3), 515–546.

Bova, F., Goldfarb, A., & Melko, R. G. (2021). Commercial applications of quantum computing. *EPJ Quantum Technology*, *8*(1), 2.

Brassard, G., Chuang, I., Lloyd, S., & Monroe, C. (1998). Quantum computing. *Proceedings of the National Academy of Sciences*, *95*(19), 11032–11033.

Braunstein, S. L. (2000). *Quantum Computing: Where Do We Want to Go Tomorrow?* (p. 305). Wiley-VCH.

Budde, F., & Volz, D. (2019). The next big thing? Quantum computing's potential impact on chemicals. *McKinsey Article*, *1*, 1–7.

Denning, D. E. (2019). Is quantum computing a cybersecurity threat? Although quantum computers currently don't have enough processing power to break encryption keys, future versions might. *American Scientist*, *107*(2), 83–86.

Feynman, R. P. (1982). Simulating physics with computers. *International Journal of Theoretical Physics*, *21*(6), 467–488.

Feynman, R. P. (1996). *Lectures in Computation*, eds. A. J. G. Hey and R. W. Allen. Addison-Wesley Publishing Company, Inc.

Gil, D., Mantas, J., Sutor, R., Kesterson-Townes, L., Flöther, F., & Schnabel, C. (2018). Coming soon to your business: Quantum computing. *IBM Institute for Business Value*. Available from: www.ibm.com/downloads/cas/OV1V0NLX. [dl: 29.07.2020].

Investors Can Now Participate in the Quantum Computing Revolution. (2016). Available from: www.equities.com/news/investors-can-now-participate-in-the-quantum-computing-revolution.

Knapp, A. (2018). Congress Just Passed a Bill to Accelerate Quantum Computing. Here's What It Does. *Forbes*.

Krelina, M. (2021). Quantum technology for military applications. *EPJ Quantum Technology*, *8*(1), 24.

Kühn, M., Zanker, S., Deglmann, P., Marthaler, M., & Weiß, H. (2019). Accuracy and resource estimations for quantum chemistry on a near-term quantum computer. *Journal of Chemical Theory and Computation*, *15*(9), 4764–4780.

Ménard, A., Ostojic, I., Patel, M., & Volz, D. (2020). A game plan for quantum computing. Available from: www.mckinsey.com/business-functions/mckinsey-digital/our-insights/a-game-plan-for-quantum-computing.

Michael, C. (2018). The business of quantum computing. *Association for Computing Machinery*, *61*(10), 20–22.

O'brien, J. L. (2007). Optical quantum computing. *Science*, *318*(5856), 1567–1570.

Palmer, J. (2012). Quantum computing: Is it possible, and should you care? Available from: www.bbc.co.uk/news/science-environment-17688257.

Parsons, D. F. (2011). Possible medical and biomedical uses of quantum computing. *Neuroquantology*, *9*(3).

Preskill, J. (2018). Quantum computing in the NISQ era and beyond. *Quantum*, *2*, 79.

Quantum Computing Market Forecast 2017–2022. (2016). Available from: www.marketresearchmedia.com/?p=850.

Rieffel, E., & Polak, W. (2000). An introduction to quantum computing for non-physicists. *ACM Computing Surveys (CSUR)*, *32*(3), 300–335.

Stantchev, R. I., Yu, X., Blu, T., & Pickwell-MacPherson, E. (2020). Real-time terahertz imaging with a single-pixel detector. *Nature Communications*, *11*(1), 1–8.

Thomson, A. (2019). *Google Says Quantum Computer Beat 10,000-Year Task in Minutes*. Bloomberg.

Van Meter, R., & Horsman, C. (2013). A blueprint for building a quantum computer. *Communications of the ACM*, *56*(10), 84–93.

Vijayalakshmi, M. (2019, October). Quantum computing in banking. *Indian Banker*, *VII*(3).

Index

A

accountability, 100
adaptability, 76
AI applications in agriculture, 30
AI applications in cybersecurity, 29
AI applications in finance, 30
AI applications in healthcare, 28
artificial intelligence, 84

B

behavioural theory, 65
busyness, 56

C

colour change, 11
command, 97
compositional grammar, 11
contingency theory, 64
control, 97
creativity, 77
Crossing the Rubicon, 52

D

diagrammatic quantum theory, 2–3
DisCoCat algorithm, 15
DisCoCirc (Distributional Composition Circuit-Shaped), 22
distributional words, 14
DNA (deoxyribonucleic acid), 39

E

effects, 4
empathy, 101
ENIAC (Electronic Numerical Integrator & Computer), 41
execution and planning of a vision, 68

F

flexible, 96
fun and value seeking, 96

G

great man theory, 63

H

human touch, 78

I

improvisation, 96
incentive, 101
integrity, 96
intellectual organization, 74

L

leadership, 48

M

machine learning models, 76
model-T leadership, 50
MVP (minimum viable product), 82

N

National Quantum Initiative Act (NQIA), 113
NISQ (Noisy Intermediate Scale Quantum Computing), 105
no direct cycles, 6
numbers, 4

P

parallel composition, 5
participative, 96
passion, 68
permanent transformation mode, 55
plurality, 96
positivity, 67
postmodern age, 49
process theory, 3–4

Q

Q-sim, 113
quantum AI, 100
quantum applications in cybersecurity, 29
quantum applications in finance, 30
quantum applications in healthcare, 28
quantum applications in logistics, 31
quantum computing, 26
quantum ghost imaging, 114
quantum illuminations, 114
quantum I-S-T-A-R, 114
quantum management, 95
Quantum Natural Language Processing, 2
quantum physics, 36
quantum radar, 114
quantum superiority, 42
quantum to business, 106

R

reflection, 68

S

self-organizing, 96
sequential composition, 5
SIEM, 29
situational theory, 64
spider fusion, 11
states, 4
strategic decision-making, 83
strategy, 82
Sycamore, 109
Sycamore quantum computer, 42

T

team building, 67
togetherness, 101
trait theory, 64

U

unbreakable encryption, 43

Z

ZX calculus, 8–11
ZX diagrams, 9